Solar Energy Pocket Reference

David Thorpe

T0174056

Routledge
Taylor & Francis Group
LONDON AND NEW YORK

from Routledge

First published 2018
by Routledge
2 Park Square, Milton Park, Abingdon, Oxon, OX14 4RN

and by Routledge
711 Third Avenue, New York, NY 10017

Routledge is an imprint of the Taylor & Francis Group, an informa business

© 2018 Taylor & Francis

Library of Congress Cataloging-in-Publication Data
Names: Thorpe, Dave, 1954– author.
Title: Solar energy pocket reference / David Thorpe.
Description: Abingdon, Oxon; New York, NY: Routledge, 2018. | Includes index.
Identifiers: LCCN 2017024891 | ISBN 9781138501201 (hbk) | ISBN 9781138806337 (pbk) | ISBN 9781315751764 (ebk)
Subjects: LCSH: Solar energy—Handbooks, manuals, etc.
Classification: LCC TJ810 .T5685 2018 | DDC 621.31/244—dc23
LC record available at https://lccn.loc.gov/2017024891

ISBN: 978-1-138-50120-1 (hbk)
ISBN: 978-1-138-80633-7 (pbk)
ISBN: 978-1-315-75176-4 (ebk)

Typeset in Goudy
by Cenveo Publisher Services

Solar Energy Pocket Reference

This handy pocket reference provides a wealth of practical information relating to solar energy and solar energy technologies. Topics covered include solar radiation and its detailed measurement, the emissivity and absorption properties of materials, solar thermal energy collection and storage, photovoltaics (both at all scales), solar cooling and the use of solar energy for desalination and drying. The book also includes conversion factors, standards and constants and is peppered throughout with helpful illustrations, equations and explanations, as well as a chapter making the business case for solar power.

Anyone with an interest in solar energy, including energy professionals and consultants, engineers, architects, academic researchers and students, will find a host of answers in this book – a practical assimilation of fundamentals, data, technologies and guidelines for application.

David Thorpe is a lecturer on one planet living at the University of Wales Trinity Saint David, a consultant on renewable energy and sustainable building, and author of several academic books and numerous articles. He is also Founder/Patron of One Planet Council.

To H.A. for bringing sunshine into my life

Contents

List of figures and tables

Figures

Tables

1 Sunlight

1.1 The nature of sunlight

Sunlight is electromagnetic radiation in the visible and near visible region of the electromagnetic spectrum. The visible light we see is a small subset of the electromagnetic spectrum shown in Figure 1.1 in grey. The breakdown of frequencies is as follows:

- ultraviolet C (UVC): 100–280 nm. Absorbed in the earth's upper atmosphere;
- ultraviolet B (UVB): 280–315 nm. Normally absorbed in the earth's upper atmosphere by the ozone layer (but absorbed less if the ozone layer has been depleted);
- ultraviolet A (UVA): 315–400 nm. (e.g. as used in tanning and therapy for psoriasis);
- visible: 400–700 nm. The range commonly used by photo-voltaic cells to produce electricity;
- infrared: 700 nm–10^6 nm (1mm).

The frequencies of sunlight reaching sea level are between 350 and 750nm.

When incoming solar radiation is absorbed – for example, on a black surface – heat is produced. This heat can be used in solar thermal technologies and passive solar buildings.

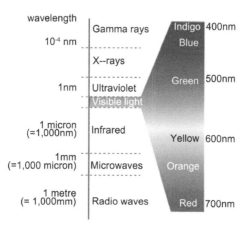

Figure 1.1 The visible light spectrum (in grey) within the electromagnetic spectrum

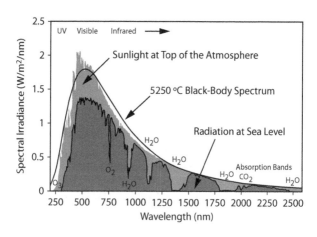

Figure 1.2 Solar radiation spectrum for direct light at both the top of the earth's atmosphere and at sea level

Source: Nick84 at https://commons.wikimedia.org/wiki/File:Solar_spectrum_ita.svg. Creative Commons Copyright

Electromagnetic radiation is characterised by 'energy packets' called photons. The amount of energy (in electron-volts) contained in a photon (E) decreases as wavelength (λ) increases, given by the Plank–Einstein equation:

$$E = \frac{hc}{\lambda} = hv \qquad [1]$$

where:

h (Planck's constant) = $6.62606957 \times 10^{-34}$ J·s
c (the speed of light in air) = 2.99792458×10^8 m/s
hc = $1.9864457 \times 10^{-25}$ J·m
v = frequency of the light.

Note that the energy per photon increases as wavelength (λ) decreases, hence UV light can damage biological cells (e.g. skin), but visible and infrared radiation is safe.

In System International (SI) units, the Planck constant is expressed in joule seconds (J·s), equal to (N·m·s). Energy passing per unit of time is power, with SI unit watt (W=J/s).

Solar *irradiance* is the rate at which the radiant energy arrives onto a unit area of a surface. It is described in watts per square metre (W/m²). The *incident* solar radiation – also sometimes referred to as *insolation* – is expressed in terms of the irradiance per time unit. If the unit is an hour (W/m²/hr) then the amount is the same figure as the irradiance.

Four characteristics of incident solar energy (insolation) are of interest:

- the spectral distribution of the light;
- the radiant power density;
- the angle at which the incident solar radiation strikes a collector surface;
- the radiant energy from the sun throughout a year or day for a particular surface.

At the atomic scale the unit of energy is the *electron-volt* (eV). This is defined as the energy required to raise an electron of electronic charge $1.602 \cdot 10^{-19}$C through 1 volt. Thus 1.000 eV = 1.602×10^{-19} J. Therefore, *hc* can be expressed in terms of eV:

$$hc = (1.99 \times 10^{-25} \, \text{J} - \text{m}) \times (1\text{eV} / 1(.602 \times 10^{-19} \, \text{J}))$$
$$= 1.24 \times 10^{-6} \, \text{eV.m}$$

Translated into micrometres (μm = 10^{-6} m):

$$hc = (1.24 \times 10^{-6} \, \text{eV.m}) \times (10^6 \, \mu\text{m/m}) = 1.24 \, eV - \mu\text{m} \qquad [2]$$

So from equation [1]:

$$E(\text{eV}) = \frac{hc}{\lambda} = 1.24 \, \text{eV.}\mu\text{m} / \lambda \qquad [3]$$

A calculator that reveals the energy of a photon for a given wavelength can be found at http://bit.ly/1euhOOb

1.2   Calculating photon flux

Calculating photon flux (ϕ) will determine the number of electrons generated and, hence, the current produced from a solar cell. It is defined as the number flux of photons per second per unit area:

$$\phi = \frac{number \ flux \ of \ photons}{\text{s.m}^2} \qquad [4]$$

The power density H is calculated by multiplying the photon flux by the energy of a single photon. Since the photon flux gives the number of photons striking a surface in a given time, multiplying ϕ by the energy of the photons comprising the photon flux gives the energy striking a surface per unit of time, which is equivalent to a power density. To determine the power density H in units of W/m², the energy of the photons must be in joules. Using equation [1] the equation is:

$$H = \phi E = \phi hc / \lambda \qquad [5]$$

Substituting from equation [3]

$$H = \phi \times (1.24 eV.\mu m) / \lambda \qquad [6]$$

where H will have the units of (W/m²).

It can be seen that the longer the wavelength, the higher the photon flux needed to yield the same power density.

1.3   Spectral irradiance

The power density distribution at a given wavelength, F, is given by:

$$F(\lambda)\Delta\lambda = \phi E \qquad [7]$$

where:

$F(\lambda)$ is the spectral irradiance in $Wm^{-2}\mu m^{-1}$
ϕ is the photon flux in the number of photons $m^{-2}sec^{-1}$
E and λ are the energy and wavelength of the photon in joules and micrometres (μm or 10^{-6} m) respectively.

Power density can also be given by:

$$F(\lambda) = \phi \, q \, \cdot \left(\frac{1.24}{\lambda(\mu m)} \right) \cdot \left(\frac{1}{\Delta\lambda(\mu m)} \right) \qquad [8]$$

where q is the charge of an electron $-1.6 \cdot 10^{-19}$C.

2 Modelling available solar radiation

Available solar radiation at a given location – latitude, longitude and height above sea level – needs to be determined as a basis for any calculations of how much energy may usefully be harvested for a project. It is usually modelled by totalling three components: direct beam shortwave radiation, sky diffuse and ground reflected radiation.[1] The available extraterrestrial radiation can be calculated using astronomical formulae and the empirical solar constant ($1,367$ Wm^{-2}).

But models vary greatly in their treatment of the atmosphere. Solar radiation is modified during its journey to the ground by scattering and absorption by air molecules, aerosols and clouds,[2] which affects wavelengths differently.[3] Cloud cover varies enormously over time and location. Factors operating at a larger scale influence the large spatial and temporal heterogeneity in local energy.[4]

Local influences include variation in air and soil temperature and moisture, vegetation type, evapotranspiration and water balance.[5] Digital elevation models can establish terrain gradient magnitude (slope) and direction (aspect), and shadows cast by nearby elevations/objects to determine the amount of beam (direct) radiation, as can calculating the local horizon angle for a given solar azimuth.[6]

Isotropic diffuse radiation models may use an average local horizon, or sky-view factor.[7] This can be reversed into a terrain-view factor and used for the reflected radiation component, normally using a constant albedo for the local surfaces.[8] The albedo is the fraction of solar energy (shortwave radiation) reflected from a surface. Anisotropic diffuse radiation models[9] may incorporate obstruction effects into the geometric description of the sky hemisphere using circumsolar and horizon bands.

The following are three different modelling systems:

- *Kumar* et al. solar model (1997): http://bit.ly/1uMwKOi. Direct, diffuse and reflected components can be integrated over daily periods
- *Solar Analyst* (Fu and Rich (2000)) uses a simple transmission model for direct radiation, and two alternative models for diffuse radiation but ignores the reflected radiation component. It can consider the effect of solar zenith angle: http://bit.ly/2tL9ghR
- *The r.sun model* (Hofierka and Ri 2002) uses Linke turbidity maps for different seasons and considers atmospheric attenuation constant over space and time when corrected for relative optical air mass. Diffuse radiation is modelled using isotropic models: http://bit.ly/1sP0XLQ and (on Europa) http://bit.ly/1w0Aopi

The following definitions may help to understand these.

2.1   Total radiant power density

This is obtained by the measured spectral distribution of irradiance (unit: W/m^3) multiplied by the wavelength range over which it was measured, then calculated over all wavelengths. The following equation calculates the total power density emitted from a light source:

$$H = \sum_i F(\lambda) \Delta \lambda \qquad [9]$$

where:

H is the total power density emitted from the light source in Wm^{-2}
$F(\lambda)$ is the spectral irradiance in units of W/m^3 and
$\Delta\lambda$ is the wavelength.

How to arrive at this value: The irradiation from the sun is not a perfectly smooth function of wavelength because atoms and molecules in both the sun itself and in the earth's atmosphere can absorb and emit at discrete wavelengths (see Figure 1.2) – that is, solar spectra contain emission and absorption lines. So the spectral width $\Delta\lambda$ is calculated from the mid-points between two adjacent wavelengths (λ_{i-1}, λ_i and λ_{i+1} etc.) as follows:

$$\Delta\lambda = \frac{\lambda_{i+1} + \lambda_i}{2} - \frac{\lambda_i + \lambda_{i-1}}{2} \tag{10}$$

$$= \frac{\lambda_{i+1} + \lambda_{i-1}}{2} \tag{11}$$

Each segment therefore contains the power:

$$H_i = \Delta\lambda \cdot F(\lambda_i) \tag{12}$$

The total power H, as in the above equation, is given by adding together all the segments.

2.2   Black-bodies

'Black-body' is the name given to the idealised surface for measuring light radiation absorption and emission. A black-body absorbs all radiation arriving on its surface and re-emits a proportion of this in a continuous frequency spectrum relative only to the body's temperature, called the Planck spectrum or *Planck's law*:

$$F(\lambda) = \frac{2\pi hc^2}{\lambda^2 (\exp\left(\dfrac{hc}{k\lambda T}\right) - 1)}$$

[13]

where:

λ is the wavelength of light (m)
T is the temperature of the black-body (K)
F is the spectral irradiance in W/m^3 and
h and c (in m/s) are constants (see above)
k, Boltzmann's constant, is 1.380×10^{-16} erg/K or 1.380×10^{-23} J/K.

The *Stefan–Boltzmann law* is used to calculate the power emitted per unit area of the surface of a black-body (J). It is directly proportional to the fourth power of its absolute temperature:

$$J^* = \sigma T^4$$

[14]

where:

J^* is the total power radiated per unit area in joule/(K/m^2)
T is the absolute temperature in degrees Kelvin and
$\sigma = 5.67 \times 10^{-8}$ W m^{-2} K^{-4} (the *Stefan–Boltzmann constant*).

To optimise power generation, seek the wavelength of peak spectral irradiance, where the most power is emitted. This is discovered using *Wien's law*, which is reached by differentiating the spectral irradiance and solving the derivative when it equals 0. Wien's law is:

$$\lambda_p = \frac{2900}{T}$$

[15]

where:

λ_p is the wavelength for the peak spectral irradiance

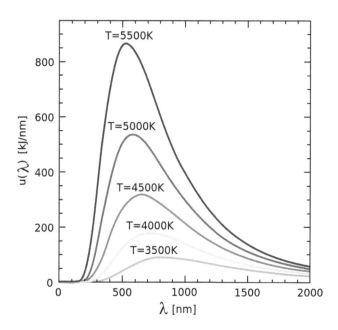

Figure 2.1 Wien's law of radiation

Source: 4C/Wikimedia Commons

2,900 is an approximation for *Wien's displacement constant*, more accurately $\lambda_p = 2.8977721(26) \times 10^{-3}$ m K and

T is the black-body's temperature in K.

2.3   Standard solar spectra

The solar spectrum changes throughout the day and with location, so standard reference spectra are deployed to permit comparison of photovoltaic devices. For terrestrial use they are defined by an internationally recognised standard known as ASTM[10] G-173-03 and International Standard ISO 9845-1:1992 as follows:

- For flat plate modules use the AM (air mass) 1.5 global spectrum. It has an integrated power of $1 kW/m^2$ ($100\ mW/cm^2$).
- For solar concentrators use the AM1.5 Direct (+circumsolar) spectrum. It combines the sun's direct beam with a component in a disk 2.5 degrees around the sun and gives a combined power density of $900 W/m^2$.

The SMARTS (simple model of the atmospheric radiative transfer of sunshine) program is used to generate the spectra. See: http://www.nrel.gov/rredc/smarts/

Further details: http://www.nrel.gov/rredc/solar_resource.html

2.4   Solar time

To determine the position of the sun it is necessary to know:

- the solar elevation (e)
- the solar azimuth (α)
- the latitude of the location (φ)
- the longitude of the location (λ)
- the equation of time (*EoT*)
- the declination (δ).

Refer to Figures 2.2–2.4. The local apparent time, or *solar time*, is calculated from the time zone standard and the so-called 'equation of time', based on the longitude. It must be expressed as an angle, known as the hour angle (ω) of the sun. This is defined as zero at solar noon and positive or negative before noon or after noon respectively. The hour angle is 15° multiplied by the fractional hours away from the solar noon when the sun is at its apogee in the sky.

The *local standard time meridian* (LSTM) for the location is calculated first. This is based on the distance in degrees of the earth's longitude from Greenwich mean time (GMT). One hour = 15° (360°/24 hours), so the equation for LSTM is:

$$LSTM = 15^{o} \cdot \Delta T_{GMT} \qquad [16]$$

Where ΔT_{GMT} is the difference between local time and GMT in hours.

2.4.1 The equation of time

In practice this correction is small (never exceeding 15 minutes), so in most cases can be neglected. If there is a need

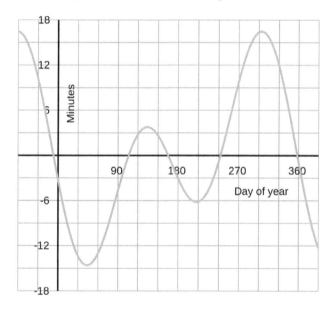

Figure 2.2 The equation of time. Local time varies compared to solar time as shown by a sundial throughout the year. When above the axis solar time will be ahead of local time and when below the axis it will appear to be behind it

Source: Wikimedia Commons, Zazou

for accuracy, we would need to adjust for the earth's 23.44° axial tilt and the eccentricity of its orbit using the *EoT*, which removes a discrepancy – due to the effect of the obliquity of the earth's rotational axis – between the actual apparent solar time, which directly tracks the motion of the sun, and the fictitious mean solar time, which is a result of the LSTM:

$$EoT = 9.8\sin(2B) - 7.53\cos(B) - 1.5\sin(B) \qquad [17]$$

where:

$$B = \frac{360}{365.24}(N - 81) \qquad [18]$$

in which N = the number of the day since the start of the year (81 is the number of days since 1 January of the spring equinox.)

We then need to compensate for the variation in solar time during one time zone. This is done by applying the LTSM to the equation of time to reach the *time correction factor* (TCF). The earth rotates 1° every four minutes, so the equation is:

$$TCF = 4(Longitude - LTSM) + EoT \qquad [19]$$

Local solar time is now found by adjusting the local time (LT) using the TCF as follows:

$$LST = LT + \frac{TCF}{60} \qquad [20]$$

2.4.2   The hour angle

We are now in a position to work out the angle of the sun at any time of day, known as the *hour angle* (ω). Since at the solar noon the angle of the sun is 0° and the earth rotates 15° for every hour passing:

$$\omega = 15°(LST - 12) \qquad [21]$$

In the morning, the angle is negative and in the afternoon it is positive.

2.4.3 *The declination angle*

Declination (δ) is the angle of an object in the sky with reference to the perpendicular to the celestial equator. The celestial equator is a projection from a point at the centre of the earth of its equator onto the celestial sphere. By convention, angles are positive to the north, negative to the south. The declination angle of the sun in the sky varies throughout the year due to the tilt of the earth on its axis of rotation as it circumnavigates the sun and gives us the seasons. The axial tilt is 23.45° and the declination angle varies plus or minus up to this full amount. At the equinoxes the

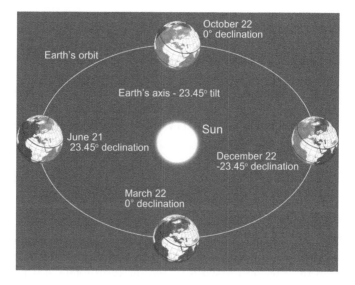

Figure 2.3 The declination of the celestial equatorial plane relative to the plane of the axis of rotation of the earth varies throughout the year due to the earth's axial tilt

value is 0°, at the 22 December solstice it is −23.45° and at the 22 June solstice it is 23.45°. It is given by the equation:

$$\delta = 23.45° \cdot sin[360 / 365 \cdot (284 + N)] \qquad [22]$$

As before, N is the number of the day in the year starting with 1 January as number 1.

2.4.4 The elevation angle

The elevation angle (ε) is the angular 'height' of the sun in the sky measured from a horizontal plane which varies according to the latitude, the time of day and the time of the year, from 0° at sunrise and sunset to 90° if the sun is directly overhead. For the design of solar power systems it is usually essential to know the maximum elevation angle for each day of the year, which occurs at solar noon and is dependent upon the latitude and declination angle (δ). It is given from the following equations:

In the Northern Hemisphere:

$\epsilon = 90° - \varphi + \delta$

In the Southern Hemisphere:

$\epsilon = 90° + \varphi - \delta$

where φ is the latitude of the location.

For a simple static solar installation it would be necessary to position the collector at the angle to be found using the equation for the equinoxes. For more sophisticated installations sun tracking is employed, so it is necessary to find ϵ for all times, which is done using the following formula:

$$\epsilon = \sin^{-1}[\sin\delta \cdot \sin\varphi + \cos\delta \cdot \cos\varphi \cdot \cos\omega] \qquad [23]$$

Where ω is the hour angle (see above).

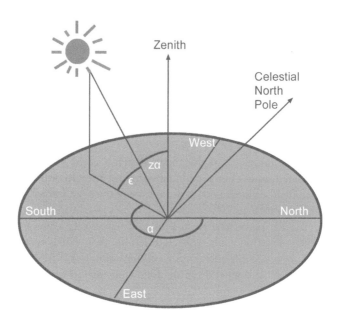

Figure 2.4 The azimuth angle (α) is measured eastward from the north. The zenith angle, also known as the incidence, ($z\alpha$) is measured from the vertical. The elevation angle (ϵ) is measured from the horizon

Source: Author

2.4.5   The azimuth angle

The azimuth angle (α) is the compass direction from which the sun's beams arrive. The usual convention is to use a north-based system in which 0° represents north, east is presented by 90°, south by 180° and west by 270°, etc. It is found by the following equation:

$$\alpha = \cos^{-1}\left[\frac{\sin\delta \cdot \sin\varphi - \cos\delta \cdot \sin\phi \cdot \cos(\omega)}{\cos\alpha}\right] \qquad [24]$$

2.5 Sun path diagrams

The procedure to determine the solar altitude and azimuth angles for a given latitude, time of year and time of day is as follows:

1. Transition the time of interest, local standard time (LST), to solar time, as above.
2. Determine the declination angle (δ) based on time of year, as above.
3. Read the solar altitude and azimuth angles from the appropriate sun path diagram on the link below. Diagrams are chosen based on latitude; linear interpolations are used for latitudes not covered. For values at southern latitudes change the sign of the solar declination.

Sunpath diagrams for each 1° of latitude for the Northern and Southern Hemisphere are available from: http://bit.ly/1mCSJQv. Figure 2.5 shows an example.

2.5.1 Shading factor

Besides being used to plot the sun's path throughout the year, the chart may be superimposed upon a photographic panorama of the site to determine if and where shading is likely to occur throughout the day and year caused by nearby objects and landscape features. On larger installations several copies of the sun path diagrams may be made so the process can be implemented at different parts of the site. An alternative method is to do it by drawing. Standing in the relevant spots on the site and looking due south (north in the Southern Hemisphere and irrespective of the orientation of the array), draw a line showing the uppermost edge of any objects that are visible on the horizon (either near or far) onto the sun path diagram. This line is called the horizon line. A shading factor is calculated as follows: if the area of the diagram that falls between sunrise and sunset is termed T,

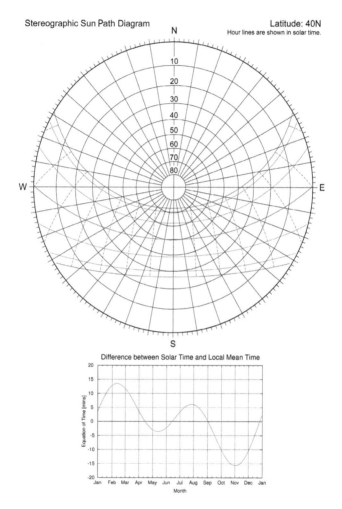

Figure 2.5 A sun path diagram

and the fraction of this area below the horizon line is termed S, then the shading factor, as a percentage of possible daylight, is:

$$SF = S / T \qquad [25]$$

The actual output of a solar installation at the site accounting for the shading factor may be calculated from the following equation:

$$kWh_a = kW_p * kWh / kW_p * SF \qquad [26]$$

where:

kWh_a = actual output over the year
kW_p= installed capacity of PV system (STC figure)
kWh/kW_p = location-determined output (either from a look-up table or calculated using the techniques in this book)
SF = shading factor.

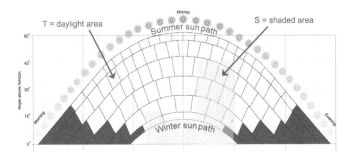

Figure 2.6 Shading (horizon line) drawn upon a sun path diagram. The total daylight area is divided into segments to aid the calculation of the total areas

2.6   Solar radiation at the earth's surface

The spectral distribution of solar radiation reaching the earth's surface at any point varies according to absorption and scattering by the atmosphere, the time of day and year, location, climate and local atmospheric conditions.

2.6.1   *Net radiation*

Net radiation (R_n), which drives weather and climate, is the balance between incoming and outgoing energy at the top of the atmosphere, being the sum of *global horizontal irradiance* (GHI) (downwards shortwave) plus upwelling shortwave plus downward and upward broadband infrared (longwave) radiation.

$$R_n = S\downarrow + S\uparrow + L\downarrow + L\uparrow \qquad [27]$$

L is measured with a pyrgeometer. R_n is measured with a radiometer. NASA publishes maps and data of net radiation in watts per square metre worldwide at: https://go.nasa.gov/2tHNBHV

2.6.2   *Air mass*

Location and time variations mean that the sun's radiation must pass through varying distances of the earth's atmosphere. This variation is measured according to the so-called air mass (M), which is defined as:

$$M = \frac{1}{\cos(\alpha)} = \frac{1}{\sin(\epsilon)} \qquad [28]$$

where α is the azimuth and ϵ is the angle of elevation. M is 1 when the sun is directly overhead (called AM1) and would be theoretically infinite at the horizon. This formula is valid at h > 10°. AM0 is the term representing daily extraterrestrial irradiance. In practice, the amount of air mass between a given point and the sun is dependent upon atmospheric pressure, the height of the point above sea level and the sun's elevation. At large zenith angles, near sunrise or sunset, the effect of atmospheric refraction can be significant.

A more accurate version, allowing for the effect of the earth's surface curvature near dawn and dusk, is:

$$M = \frac{1}{\cos(\alpha) + 0.50572(96.07995 - \alpha)^{-1.6364}} \qquad [29]$$

A calculator which determines the amount of solar radiation hitting a PV module under clear-sky conditions is available at http://bit.ly/1nqduQ6

A simple method of calculating the air mass at any given time involves measuring the shadow length (s) of a vertical pole of height (h) and using the following equation:

$$M = \sqrt{1 + \left(\frac{s}{h}\right)^2} \qquad [30]$$

M is calculated relative to the standard atmospheric pressure (P_o) at sea level, of 101.324 hectapascals (hPa) or 1013.25 millibar (mb). Site air pressure (P_s) is related to altitude (h). If no data for this are available, an approximation is given by:

$$P_s = P_o{}^{e-(000832/km)h} \qquad [31]$$

where h is height above sea level. The pressure-corrected air mass M_p, relative to $z = 0$ and sea level is then given by:

$$M_p = \frac{MP_s}{P_o} \qquad [32]$$

For larger zenith angles, approaching sunset and sunrise, refraction considerably distorts this. Therefore a more accurate algorithm for computing refracted air mass (M_r) from the apparent solar zenith angle (z) is:

$$M_r = \frac{1}{\cos z} + 0.50572 \cdot (96.07995 - z)^{-1.6364} \qquad [33]$$

(Kasten and Young[11])

The refraction-corrected air mass, M_rP, is then corrected for the air pressure at the location using the previous equation:

$$M_r P = \frac{M_r \cdot P_s}{P_o} \qquad [34]$$

An alternative method to calculate the amount of direct irradiation (I_{dir}) on a plane perpendicular to the sun at any given moment in the day (a tracking module) is to use the equation:

$$I_{dir} = 1.353 \cdot \left[\left(1 - ah \right) 0.7^{AM^{0.678}} \right) + ah] \qquad [35]$$

where:

1.353 kW/m^2 is the solar constant
0.7 is the fraction of total solar energy at the outer edge of the earth's atmosphere that reaches the surface
0.678 is an empirically observed factor accounting for variations in the atmosphere
a = 0.14km^{-1}
h = height in km above sea level.

2.6.3   Global horizontal irradiance (GHI)

The most common paramater for calculations at a location is GHI. This is the total solar flux incident on a horizontal surface, implying a 2π steradian field of view. It is the sum of the direct normal irradiance (DNI) and all other radiation coming from the sky dome.

$$GHI = DNI^* \cos(\theta) + DHI \qquad [36]$$

where: θ = solar zenith angle and DHI = direct horizontal irradiation (measured by a properly shaded pyranometer).

This is then integrated via a data logger into hourly or minute irradiation.

With concentrating solar power systems the DNI is the only irradiance measure needed.

For non-concentrating flat plate collectors global tilted irradiance (GTI) is required. GTI is specified by tilt in degrees from horizontal and orientation, eastward from true North – for example, GTI (20°, 180°). A generic formula for calculating GTI is:

$$GTI = DNI \cdot \cos(z_T) + DHI \cdot \left(\frac{1 + \cos(T)}{2}\right) + GHI \cdot \rho \cdot \left(\frac{1 - \cos(T)}{2}\right)$$
$$[37]$$

Where z_T is the angle of incidence relative to the tilted surface, T is the tilt angle and ρ is the average albedo of the nearby ground, for the measurement of which a pyranometer is required.

If the collector plane is tilted away from the horizontal, then add the reflected irradiance (I_r) to this. See section 2.7 for more on this.

If the collector is tracking, then the sum of the irradiance is called global normal irradiance (GNI) and is defined as:

$$GNI = DNI + DTI + I_r \qquad [38]$$

2.6.4 The clearness index

The clearness index indicates the overall transmittance of the atmosphere at the site. A clearness index can be derived by normalising the global (K_t), direct (K_n) and diffuse (K_d) irradiance measurements by the extraterrestrial solar radiation at the top of the atmosphere on either a horizontal surface or normal to the sun. By definition:

$$K_t = K_n + K_d \tag{39}$$

It is a ratio of measured irradiation in a locale relative to the extraterrestrial irradiation Io/AM0. The total global hemispherical component on a horizontal surface clearness index (K_t) is:

$$K_t = \frac{GHI}{I_o \cdot \cos(z)} \tag{40}$$

The diffuse hemispherical component on a horizontal surface clearness index (K_d) is:

$$K_d = \frac{DHI}{I_o \cdot \cos(z)} \tag{41}$$

For $K_t \to 1$: atmosphere is clear. For $K_t \to 0$: atmosphere is cloudy. This measure incorporates both light scattering and light absorption. K_t values depend on the location and the time of year; they are usually between 0.3 (for very overcast climates) and 0.8 (for very sunny locations).

2.6.5   The clear-sky index

This is an alternate indicator for the way that the atmosphere attenuates light on an hour to hour or day to day basis, defined as:

$$k_c = \text{measured/calculated clear sky}$$

$1 - k_c$ is a good indicator of the degree of 'cloudiness' in the sky. All the core research for the empirical calculations used in software like TRNSYS, Energy+ and SAM was based on k_T.

It is possible to use the power output from one system to estimate the power output of another nearby system, but the process is not necessarily straightforward. A modified clear-sky index for photovoltaics can be deployed which uses the ratio of the instantaneous PV power output to the instantaneous theoretical clear-sky power output derived from a clear-sky radiation model and PV system simulation routine. This definition performs better than previous clear-sky indices when both PV systems' characteristics are known and the two PV systems have similar orientations.[12]

2.7   Tilted surfaces

2.7.1   *Solar incidence angles*

For a horizontal surface the incidence angle (θ) of a direct sunlight beam is 90 – e (the elevation angle – $z\alpha$ in Figure 2.4). For a tilted surface the equation for the incidence angle (θ) must include the tilt angle (s), azimuth angle and the azimuth of the tilted surface (γ) With respect to south = 0 this is:

$$
\begin{aligned}
\cos(\theta) = &\left(\sin\delta\cdot\sin\varphi\cdot\cos s\right) - \left(\sin\delta\cdot\cos\varphi\cdot\sin s\cdot\cos\gamma\right) \\
&+ \left(\cos\delta\cdot\cos\varphi\cdot\cos s\cdot\cos\omega\right) \\
&+ \left(\cos\delta\cdot\sin\varphi\cdot\sin s\cdot\cos\gamma\cdot\cos\omega\right) \\
&+ \left(\cos\delta\cdot\sin s\cdot\sin\gamma\cdot\sin\omega\right)
\end{aligned}
\tag{42}
$$

The incident angle is the inverse trigonometric function of this.

2.7.2   *Irradiance components*

On a tilted plane, such as a panel or roof, the total irradiance (I_{tilt}) consists of three components, direct (I_{dir}), diffuse (I_{dif}) and reflected irradiance (I_r):

$$
\begin{aligned}
I_{tilt} &= I_{dir} + I_{dif} + I_r \\
&\approx (I_{dir}\cdot 1.1) + (I_{dir}\cdot 0.2)
\end{aligned}
\tag{43}
$$

(Average ground reflection is about 20 per cent of global irradiance. Otherwise use the values in Table 2.1.)

Table 2.1 The albedo of some natural and artificial surfaces

Surface	Albedo (%)
Forests	10–20
Soil	14
Water bodies (varies with sun altitude)	10–60
Dry grass	20
Grass	25–30
Crops, grasslands	26
Dry leaves	30
Fresh snow	75–90
Earth roads	4
Asphalt (blacktop)	5–10
Dark roof	8–18
Bituminous gravel roof	13
Concrete, dry	17–27
Crushed rock	20
Red brick	27
Light roof	35–50
Light brick	60

The total irradiance can also be arrived at from multiplying the normal irradiation by the cosine of the incidence angle (θ) and adding the diffuse irradiation and reflected radiation.

$$I_{tilt} = I_{dir} \cdot \cos(\theta) + I_{dif} + I_r \qquad [44]$$

Irradiance is generally measured in J/m^2 and is represented by the symbol H (defined in equation 9 above).

Diffuse solar irradiation (I_{dif}) is that scattered by aerosols and dust particles in the atmosphere. It is primarily in the blue part of the spectrum, giving the sky its colour. On a clear day, the energy flux in the blue is about 10 per cent of the total solar irradiance.

Liu and Jordan's[13] isotropic diffuse algorithm to compute monthly average radiation in the plane of the collector, *HT*:

$$\bar{H}_T = \bar{H}_b \bar{R}_b + \bar{H}_d \left[\frac{1+\cos\beta}{2} \right] + \bar{H}\rho_g \left[\frac{1-\cos\beta}{2} \right] \qquad [45]$$

where:

H_b is the product of monthly average beam radiation

R_b depends on collector orientation, site latitude and time of year

H_d represents the contribution of monthly average diffuse radiation, which depends on the angle of the collector, β and

ρ_g is a value for the reflected radiation from the ground in front of the collector, and depends on the slope of the collector and on ground reflectivity. This latter value is assumed to be equal to 0.2 when the monthly average temperature is above 0°C and 0.7 when it is below −5°C, and to vary linearly with temperature between these two thresholds.

Monthly average daily diffuse radiation is calculated from global radiation through the following formulae:

a)   for values of the sunset hour angle ω_s less than 81.4°:

$$\frac{\bar{H}_d}{\bar{H}} = 1.391 - 3.560\bar{K}_T + 4.189\bar{K}_T^2 - 2.137K_T^3 \qquad [46]$$

b)   for values of the sunset hour angle ω_s greater than 81.4°:

$$\frac{\bar{H}_d}{\bar{H}} = 1.311 - 3.022\bar{K}_T + 3.427\bar{K}_T^2 - 1.821K_T^3 \qquad [47]$$

The monthly average daily beam radiation H_b is simply computed from:

$$\bar{H}_b = \bar{H} - \bar{H}_d \qquad [48]$$

The UV component of the total radiation can be estimated as follows:

$$\frac{I_{UV.h}}{I_T} = 0.14315K_t^2 - 0.20445K_t + 0.135544$$

$$\frac{I_{UV.b}}{I_{T.b}} = 0.688e^{-0.575m} = 0.688\,exp\left[\frac{-0.575}{sin\alpha}\right]$$

[49]

where:

I_t = global horizontal total solar radiation
$I_{t.b}$ = beam component of total solar radiation
$I_{UV.b}$ = beam component of UV radiation
$I_{UV.h}$ = global horizontal UV radiation
K_t = clearness index
m = air mass
α = solar altitude angle.

Notes

1 Perez, R., Scott, R., Arbogast, C., and Scott, J., 'An anisotropic hourly diffuse radiation model for sloping surfaces: description, performance validation, site dependency evaluation', *Solar Energy*, 36, 6 (1986).

2 Hofierka, J. and Ri, M., 'The solar radiation model for open source GIS: implementation and applications', Proceedings of the Open Source GIS – GRASS Users Conference, Trento, Italy, 11–13 September 2002.

3 Bird, R., and Riordan, C. 'Simple solar spectral model for direct and diffuse irradiance on horizontal and tilted planes at the earth's surface for cloudless atmospheres', *Journal of Climate and Applied Meteorology*, 25, 87–97 (1986), and Wald, L. 'SODA: a project for the integration and exploitation of networked solar radiation databases', European Geophysical Society Meeting, XXV General Assembly, Nice, France. 25–29 April 2000.

4 Gates, D.M., *Biophysical Ecology*, Springer-Verlang, New York (1980), and Rich, P.M., and Fu, P., 'Enlightenment for mapping

systems: solar radiation models look to the sun for answers', *Resource Magazine*, 6 (2), 7/8 (2000).

5 Allen, R.G., Pereira, L.S., Raes, D., and Smith, M., 'Crop evapotranspiration: guidelines for computing crop water requirements – FAO irrigation and drainage paper 56', Food and Agriculture Organization (FAO) of the United Nations Rome (1998); and Gates (1980).

6 Dozier, J., and Frew, J., 'Rapid calculation of terrain parameters for radiation modeling from digital elevation data', *IEEE Transactions on Geoscience and Remote Sensing*, 28, 963–9 (1990).

7 Iqbal, M., *An Introduction to Solar Radiation*, Academic Press, Cambridge, MA (1983).

8 Gates (1980) and Allen *et al.* (1998).

9 Perez *et al.* (1986).

10 American Society for Testing and Materials.

11 Kasten, F., and Young, A.T., 'Solar position algorithm for solar radiation applications', *Solar Energy*, 76 (5), 577–89 (1989).

12 Engerer, N.A., and Mills, F.P., 'KPV: a clear-sky index for photovoltaics', *Solar Energy*, 105 (July 2014), 679–93.

13 Liu, B.Y.H., and Jordan, R.C., 'The interrelationship and characteristic distribution of direct, diffuse and total solar radiation', *Solar Energy*, 4 (3), 1–19 (1960).

3 Emissivity and absorption of materials

All materials have the capacity to absorb and emit the energy they receive from the sun. Designers of solar technologies wish to maximise the proportion of incident solar energy absorbed and control how much is emitted. Absorption is a measure of how much incident radiation is received by a material. The emissivity of the surface of a material is a measure of its effectiveness in emitting energy.

The emitted energy is in the form of long wave or thermal radiation. In scientific terms:

- *Absorption* refers to the ability of a material to absorb solar radiation (approximately the wavelengths 400nm–10µm (UV-A, visible and near infrared)).
- *Emissivity* or *emittance* refers to the ability of a material to emit infrared radiation (from materials with temperatures between −40°C to 100°C, of approximately the wavelengths 10µm–100µm).

Since energy cannot be lost or destroyed, any radiation not absorbed into a material will be reflected. The radiation that is absorbed will have the effect of heating the material, which then itself will emit heat (infrared radiation) in order to reach the same temperature as its surroundings. If a material absorbs a

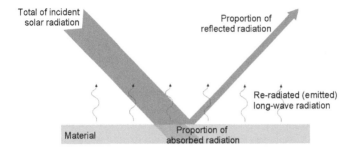

Figure 3.1 Absorption and emittance of incident solar energy in a material

great deal of solar radiation it is said to have a high absorption coefficient. If it emits much radiation it is said to have a high emissivity coefficient. The converse is true for low absorption or emissivity rates.

- An absorption coefficient of 1 would mean that all of the solar spectrum radiation was absorbed and of 0 would mean that none of it was absorbed and all of it was reflected. By definition a true 'black-body' surface has an absorption coefficient of 1.
- An emissivity coefficient of 1 would mean that all of the incident radiation was re-emitted as infrared radiation and of 0 would mean that none of it was.

(In practice these extremes are not reached.)

3.1   Selective surfaces

The need for spectral selectivity is dependent on the intensity of the incoming solar radiation and the temperature at which the surface is to operate. A black absorber and a white reflector will absorb and reflect all energy incident on them, regardless of wavelength. Selective absorbers will absorb only in the spectral region of the solar radiation and be transparent in the thermal infrared. Selective reflectors will reflect only in the thermal infrared and be transparent in the spectral range, so the solar radiation will travel straight through and be absorbed in a black absorber layer put beneath that layer.

Selective surfaces therefore need to have the appropriate spectral profile, plus resistance to the operating temperature, be efficient and economic to manufacture and scale, and long-lasting. No single material can do this, so the tandem approach has been developed. These combinations of materials in layers are called absorber-reflector tandems.

Figure 3.2 Roof on a building in Los Angeles, California, given a highly reflective surface to keep out unwanted solar heat

In the manufacture of solar technologies, surface layers with a high degree of spectral selectivity are applied to materials to achieve the desired levels of emittance and absorption. Typical values for a selective surface coefficient might be 0.90 for absorption and 0.10 for emissivity. They can range from 0.8/0.3 for paints on metal to 0.96/0.05 for commercial surfaces. Thermal emissivities as low as 0.02 have been obtained in laboratories. In large-scale commercial applications these layers are applied at a nano-scale.

The exterior surfaces of solar thermal absorber tubes need to provide higher absorbance, lower emittance and resistance to atmospheric oxidation at elevated temperatures. The coatings are multilayered, comprising solar absorbent layers topped with antireflective layers of refractory metal or metalloid oxides (such as titania and silica) with differing indices of refraction.

At least one layer of a noble metal such as platinum may be included. The absorbent layers may include heat-resistant materials including oxides of refractory metals or metalloids such as silicon. Reflective layers may comprise refractory metal silicides or related compounds of titanium, hafnium or zirconium.

An absorber-reflector tandem is obtained by combining two surfaces, one which is highly absorbing in the solar region and another highly reflecting in the infrared – for example, by covering a base metal of high infrared reflectance with a thin highly solar absorbing coating; or by covering a thick absorbing surface with a solar transparent infrared reflecting coating.

Spectrally selective semiconductor coatings are obtained by depositing a semiconductor with a low band gap (see box below for explanation) – so that it absorbs the solar radiation – on a highly infrared reflecting metal substrate.

3.1.1 Band gaps

In a given material the highest range of electron energies in which electrons are normally present at absolute zero temperature is called the *valence band*, while the *conduction band* is the

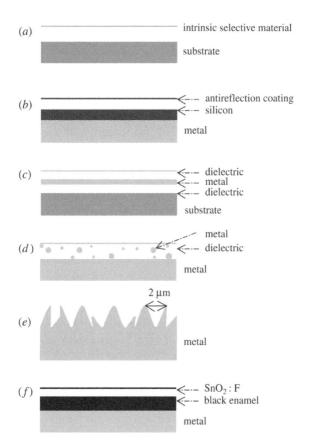

Figure 3.3 A schematic of six different coatings and surface treatments for selective absorption of solar energy:

3a: a material which has appropriate intrinsic optical properties which are ideal for solar control materials

3b: a semiconductor metal tandem giving the required spectral selectivity by the absorption of short-wavelength radiation in a semiconductor which has a band gap of 0.6 eV. It also has low-Etherm owing to the metal also on the surface. However, useful semiconductors usually have large refractive indexes, which result in high reflection losses and create a need for an extra antireflection coating that is effective in the solar range

3c: multi-layer coatings consisting of dielectric/metal/dielectric (D/M/D) that make efficient selective solar absorbers. Coatings with Al_2O_3 and Mo as the dielectric and metal respectively have good properties and can be produced on a large scale using vacuum coating technology

3d: a metal/dielectric-composite-metal tandem, which contains nanoparticles in a dielectric host

3e: textured metal surfaces give numerous reflections in metal dendrites which are 2000 nm apart with high absorption, but not so low emittance because relevant wavelengths are much larger than the dendrite separation

3f: A selectively solar-transmitting film on a black-body-like absorber gives good spectral selectivity

Source: Adapted from Granqvist, C.G., 'Solar energy materials', *Applied Physics A: Materials Science and Processing*, 52, 83 (1991). doi:10.1007/BF00323721. CrossRef Google Scholar, Copyright-free

nearest range of vacant electronic states. The valence band is the lower energy level and the conduction band is the higher energy level. It therefore takes energy to make an electron jump to the conduction band; this comes from the photons arriving from the sun. In semiconductor and insulating materials the two bands are separated by what is known as a 'band gap'. The degree of electrical conductivity depends on how easily electrons may flow from the valence band (where electrons are 'stuck') to the conduction band (where they may flow).

When electrons jump from the valence to the conduction band the material is conducting electricity, and the hole left behind in the valence band is available to be filled by more electrons (if in a circuit).

The band gap of a semiconductor – as used in a photovoltaic cell – determines how much energy is needed from the sun to trigger the jumping, and how much energy is generated. In other words, the band gap is the minimum energy required to send an electron out of a valence band. It is usually expressed in units of electron Volts (eV) where 'one electron' is the elementary charge, e.

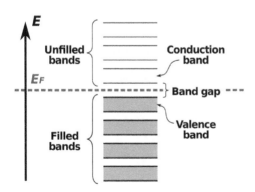

Figure 3.4 Semiconductor band structure, showing a few bands on either side of the band gap

Source: Tim Starling, Creative Commons

In physical terms, the valence band and conduction band are the bands closest to what is known as the Fermi level (the electrochemical potential for electrons). They determine the electrical conductivity of the material.

3.2   Emittance

The emission coefficient depends on the surface condition of the material, including its roughness, surface and oxide layers. Emissivity refers to the radiation of power from a surface at a certain temperature. But radiated power varies with temperature, being greater at higher temperatures (see Figure 3.5). Most solar collectors are, in practice, operated at temperatures above that used to measure the emissivity factor, sometimes causing the amount of emitted radiation to be higher than expected.

The emissivity coefficient of a material – ε – indicates the radiation of heat from a 'grey body' (E) compared with the radiation of heat from an ideal 'black body'(E_b) with an emissivity coefficient of $\varepsilon = 1$ (see above).

$$\varepsilon = \frac{E}{E_b} = \frac{1}{\sigma T^4} \int_0^\infty \varepsilon_\lambda E_{b\lambda} d\lambda \qquad [50]$$

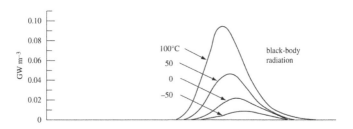

Figure 3.5 Spectra showing black-body radiation for four different temperatures

where:

σ is the Stefan-Boltzmann constant (5.670×10^{-8}W/m^2 – K^4)
T is the surface temperature (absolute scale) and
λ refers to the wavelength, since emissivity varies with the wavelength: $\varepsilon = \varepsilon(\lambda)$.

Since energy cannot be destroyed, the incident radiation balance is:

$$\alpha + \tau + \rho = 1 \qquad [51]$$

where:

I is the total surface irradiation

$\alpha(\text{absorptivity}) \equiv \dfrac{I_{absorbed}}{I}$

$\tau(\text{transmissivity}) \equiv \dfrac{I_{trans}}{I}$ and

$\rho(\text{reflectivity}) \equiv \dfrac{I_{refl}}{I}$.

The *irradiance* (j^*) has dimensions of energy flux (energy per time per area). Its SI units of measure are joules per second per square metre or, equivalently, watts per square metre.

Radiance (L) (in watts per square metre per steradian[1]) is given by:

$$L = \frac{j^*}{\pi} = \frac{\sigma}{\pi} \cdot T^4 \qquad [52]$$

To find the total power, P, radiated from an object, multiply by its surface area, A:

$$P = A \cdot j^* = A \cdot \varepsilon \cdot \sigma \cdot T^4 \qquad [53]$$

Table 3.1 The relationship between colour and absorption factors

Surface colour	Absorption factor: α (approximated)
White smooth surfaces	0.25–0.40
Grey to dark grey	0.40–0.50
Green, red and brown	0.50–0.70
Dark brown to blue	0.70–0.80
Dark blue to black	0.80–0.90

3.3   Absorptance

The absorption spectrum is the fraction of incident radiation absorbed by a material over a range of radiation frequencies. The wider the spectrum, the more radiant energy is absorbed.

In general, the solar energy absorbed – for example, by a wall or roof – can be approximated according to the surface colour, as shown in Table 3.1.

As in Figure 3.1, a fraction of the solar irradiance incident on a given surface is absorbed and is converted into heat – and in the case of a photovoltaic cell into electricity – and the remaining fraction is reflected and lost.

The amount of solar energy absorbed also depends upon the angle at which it arrives, as shown in Table 3.2.

An understanding of the factors determining the absorption factor is important in PV applications because it is one of the major parameters determining the solar cell temperature under operational conditions, thereby influencing the cell efficiency and the electrical yield. It is also relevant to so-called photovoltaic/thermal (PVT) combi-panels, where the energy not captured by the photoelectric effect is collected and used for heating water and/or space.

Spectral absorptance (α) is calculated using Kirchoff's law, which expresses it in terms of total reflectance ρ (λ, θ) for opaque materials:

Table 3.2 Angular variation of the absorptivity of lampblack paint

Incidence angle i(°)	Absorptance a(i)
0–30	0.96
30–40	0.95
40–50	0.93
50–60	0.91
60–70	0.88
70–80	0.81
80–90	0.66

$$\alpha(\lambda,\theta) = 1 - \rho(\lambda,\theta)$$

and

$$\varepsilon(\lambda,T) = \alpha(\lambda,T) \qquad [54]$$

where:

$\rho(\lambda,\theta)$ is the sum of both collimated[2] and diffuse reflectance
λ is the wavelength
θ is the incidence angle
T is the given temperature
ε is the emissivity coefficient.

The *solar absorptance* is obtained by weighting the spectral absorptance with the spectral solar irradiance and, for a given angle of incidence θ, it can be obtained by integrating over the wavelength dependent solar spectrum, $G(\lambda)$:

$$\alpha_{sol}(\theta) = \frac{\int_{\lambda_1}^{\lambda_2} \left[1 - R(\lambda,\theta)\right] G(\lambda) d\lambda}{\int_{\lambda_1}^{\lambda_2} G(\lambda) d\lambda} \qquad [55]$$

where λ_1, λ_2 denote the lower and upper solar wavelengths respectively.

3.4   Absorption and emissivity of selective surfaces

Table 3.3 Absorption and emissivity values of selective surfaces

Product	Absorption	Emissivity	Ratio	Comments
Black crystal	0.92–0.98	0.08–0.25	5.76	Thermafin's black crystal selective surface coating
Copper treated with $NaClO_2$ and NaOH	0.87	0.13	6.69	
Copper, aluminium or nickel plate with CuO coating	0.08–0.93	0.09–0.21	3.37	
Metal, plated black chrome	0.87	0.09	9.70	
Metal, plated black sulphide	0.92	0.10	9.20	
Metal, plated nickel oxide	0.92	0.08	11.00	Stainless steel heated until the nickel oxidises
Solchrome	0.94–0.98	0.10–0.14	8.00	Made by Solchrome Systems India Ltd
Solec LO/MIT selective surface paint	0.21–0.26	0.15–0.19	1.38	LO/MIT I/II products are low emissivity, non-thickness-dependent coatings
Solec SOLKOTE selective surface paint	0.88–0.94	0.28–0.49	2.36	SOLKOTE HI/SORB-II is an optical coating specifically formulated for solar thermal applications
White paint	0.23–0.49			

3.5   Absorbed solar radiation by material

The emissivity and absorption coefficients for some common materials can be found in Table 3.4. Note that the coefficient for some products varies with the temperature. As a guideline, the emittances are based on a temperature of 300°K. It's vital to select a surface that is optimised for the actual operating temperature because small errors in measured ρ (reflectance) can lead to large errors in small values of ε.

Table 3.4 Absorption and emissivity factors of selected materials, and their ratio

Material	Absorption	Emissivity	Ratio
Aluminium oxide paint	0.09	0.92	0.1
Aluminium, anodised	0.14–0.15	0.77–0.84	0.17
Aluminium foil	0.15	0.05	3
Aluminium, highly polished	0.039–0.057	0.09	
Aluminium paint (bright)	0.30–0.50	0.40–0.60	0.8
Aluminium, polished	0.09	0.03	3
Aluminium, highly polished		0.04–0.06	3
Antimony, polished	0.3	0.28–0.31 1	
Asbestos board	0.96	0.83	1.157
Asphalt	0.93	0.91 (new), 0.82 (old)	1.021 (1.13)
Barium sulphate with polyvinyl alcohol	0.06	0.88	0.07
Basalt		0.72	
Beryllium		0.18	
Beryllium, anodised		0.90	
Biphenyl-white solid	0.23	0.86	0.27
Bitumen-covered roofing sheet, brown		0.87	
Black-body matt			1
Black crystal	0.92–0.98	0.08–0.25	5.76
Black enamel paint		0.80	
Black epoxy paint		0.89	
Black lacquer on iron		0.875	

(*Continued*)

Material	Absorption	Emissivity	Ratio
Black paint (average)	0.96	0.86	1.12
Black Parson optical		0.95	
Black plastic	0.96	0.87	1.1
Black silicate	0.96	0.89	1.08
Black silicone paint		0.93	
Brick, fireclay	0.75	glazed: 0.35	
Brick, red (rough)	0.65	0.93	0.68
Carbon black paint	0.96	0.88	1.09
Carbon filament		0.77	
Carbon pressed filled surface		0.98	
Cast iron, newly turned		0.44	
Cast iron, turned and heated		0.60–0.70	
Catalac white paint	0.24	0.9	0.27
Chemglaze black paint z3o6	0.96	0.91	1.05
Chromium		0.08–0.26	3
Chromium polished		0.058	
Concrete	0.6	0.85–0.88	0.68
Concrete and stone, dark	0.65–0.80	0.85–0.95	0.81
Concrete, rough		0.94	
Copper, heated and covered with thick oxide layer	0.78	0.64	
Copper nickel alloy, polished		0.059	
Copper, polished	0.18	0.023 –0.052	4.5
Dow corning white paint dc-007	0.19	0.88	0.22
Dull brass, copper, galvanised steel, aluminium	0.40–0.65	0.20–0.30	2.1
Dupont lucite actylic lacquer	0.35	0.9	0.39
Ebanol c black	0.97	0.73	1.33
Ebanol c black-384 esh* uv	0.97	0.75	1.29
Flat black paint	0.97–0.99	0.97–0.99	1

(*Continued*)

Material	Absorption	Emissivity	Ratio
Galvanised metal, new	0.65	0.13	5
Galvanised metal, weathered	0.8	0.28	2.9
Glass, smooth		0.92–0.94	
Gold, not polished		0.47	
Gold, highly polished		0.02–0.04	3
Granite	0.45	0.55	0.818
Graphite		0.84	
Gypsum			0.85
Ice, rough		0.97	
Iron and steel, strongly oxidised		0.95	3
Iron, polished	0.64	0.14–0.38	
Iron, dark grey surface	0.31		
Iron, plate rusted red	0.92	0.61	
Iron, rough ingot		0.87–0.95	
Lead, oxidised	0.79	0.43	0.14
Lead, pure unoxidised		0.057–0.075	
Light coloured paints, firebrick, clay, glass	0.04–0.40	0.9	0.24
Limestone	0.35–0.50	0.90–0.93	0.464
Limewash		0.91	
Magnesium oxide	0.08	0.20–0.55	0.40–0.14
Magnesium oxide paint	0.09	0.9	0.1
Marble, white	0.44	0.95	0.46
Masonry, plastered	0.93		
Mercury liquid		0.1	
Metal, plated black chrome	0.87	0.09	9.7
Metal, plated black sulphide	0.92	0.1	9.2
Metal, plated cobalt oxide	0.93	0.3	3.1
Metal, plated nickel oxide	0.92	0.08	11
Molybdenum, polished		0.05–0.18	
Nichrome wire, bright		0.65–0.79	
Nickel, elctroplated		0.03	
Nickel, oxidised		0.59–0.86	
Nickel, polished		0.072	
Oak, planed	0.89		
Opal glass	0.28	0.87	0.32
Plaster		0.98	

(*Continued*)

Material	Absorption	Emissivity	Ratio
Plaster, rough		0.91	
Platinum, polished plate		0.054–0.104	
Polyethylene black plastic	0.94	0.92	1.01
Porcelain, glazed	0.5	0.92	0.54
Pyramil black on beryllium copper	0.92	0.72	1.28
Quartz glass		0.93	
Roofing paper		0.91	
Rubber (natural) hard		0.91	
Rubber (natural) soft	0.65	0.86	0.756
Sand/sandstone	0.62–0.73	0.76	0.88
Silicon carbide		0.83–0.96	
Silver, highly polished		0.02–0.03	3
Slate		0.87	
Snow, fine particles fresh	0.13	0.82	0.16
Snow, ice granules	0.33	0.89	0.37
Stainless steel, polished	0.76–0.80	0.075	1.04
Stainless steel, weathered		0.85	
Tile (red clay)	0.64	0.97	0.66
Tin unoxidised		0.04	
Titanium oxide white paint with methyl silicone	0.2	0.9	0.22
Titanium oxide white paint with potassium silicate	0.17	0.92	0.18
Titanium, polished		0.19	
Tungsten, aged filament		0.032–0.35	
Tungsten, polished		0.04	
Water		0.95–0.963	
White paint (average)	0.39	0.89	0.43
Wood, beech, planed		0.935	
Wood, oak, planed		0.885	
Wood, pine		0.95	
Zinc orthotitanate with potassium silicate	0.13	0.92	0.14
Zinc oxide with sodium silicate	0.15	0.92	0.16
Zinc, polished	0.045		
Zinc, tarnished	0.25		
Zirconium oxide with glass resin	0.23	0.88	0.26

3.6 Spectral absorption of solar radiation in water

This may be of relevance for solar ponds. A solar pond is a pool of saltwater used to collect and store solar thermal energy in hot, predominantly dry, countries. Low-salinity water floats naturally to the top, with the denser, high-salinity water at the bottom. The addition of a saturating quantity of salt greatly reduces heat loss and allows for the high-salinity water to rise up to 85°C. The heat is drawn off through a heat exchanger to power a heat engine and supply electricity. A large thermal mass (salt store) allows power to be generated after sunset and through the night.

According to the second law of thermodynamics, the maximum theoretical efficiency of a cycle that uses heat from a high temperature reservoir at 80°C and has a lower temperature of 20°C is $1 - (273 + 20)/(273 + 80) = 17\%$. The solar energy conversion in practice is typically less than 2 per cent, but the low efficiency is justified when the low-tech 'collector' – typically a large black plastic-lined pond – displaces a more expensive large-scale system.

A solar pond has three zones with the following salinity with depth:

1. surface convective zone (0.3–0.5m), with salinity <5% concentration

Table 3.5 The percentage of sunlight in the wavelength band passing through water of the indicated thickness

Wavelength	Layer depth				
(μm)	0	1cm	10cm	1m	10m
0.2–0.6	23.7	23.7	23.6	22.9	17.2
0.6–0.9	36.0	35.3	36.0	12.9	0.9
0.9–1.2	17.9	12.3	0.8	0.0	0.0
1.2 and over	22.4	1.7	0.0	0.0	0.0
Total	100.0	73.0	54.9	35.8	18.1

2. non-convective zone (1–1.5m), salinity increasing with depth
3. storage zone (1.5–2m, salt=20%).[3]

3.7 Light refraction in transparent materials

The refraction of light between materials – for example, the type of materials that are used in thermal collector covers – is given by the following equation:

$$\frac{\sin(i)}{\sin(\theta_r)} = \frac{n'_r}{n'_i} = n_r \qquad [56]$$

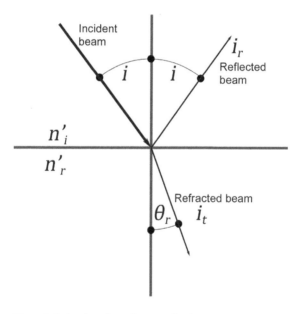

Figure 3.6 Angles of incidence and refraction

Table 3.6 Visible spectrum refractive index values, n_r, based on air

Material	Index of refraction
Air	1.00
Clean polycarbonate	1.59
Diamond	2.42
Glass (solar collector type)	1.50–1.52
Plexiglass (polymethyl methacrylate, PMMA)	1.49
Mylar (polyethylene terephthalate, PET)	1.64
Quartz	1.54
Tedlar (polyvinyl fluoride, PVF)	1.45
Teflon (polyfluoroethylenepropylene, FEP)	1.34
Water – liquid	1.33

Notes

1 The steradian (symbolised sr) is the Standard International (SI) unit of solid angular measure. In a complete sphere there are 4 pi – approximately 12.5664 – steradians. A steradian is conical in shape. The point of the cone is the centre of the sphere. The length of one side is the radius of the sphere. The curved area of the end of the cone is the square of the radius.
2 Collimated light rays are parallel, spreading minimally as they travel, not dispersing with distance. Diffuse light is reflected or refracted light arriving at many angles, of a lower intensity.
3 More information at http://bit.ly/2eCLBcf

4 Photovoltaics

4.1 Photovoltaic cells

Photovoltaic (PV) solar cells directly convert sunlight into electricity using the photovoltaic effect. PV cells are assembled into modules to build modular arrays that generate electricity in both grid-connected and off-grid applications. Commercial PV technologies include:

- *wafer-based crystalline silicon* (c-Si) (either mono-crystalline or multi-crystalline silicon): maximum efficiency of around 25 per cent. The best current commercial modules are 19–20 per cent (with a target of 23 per cent by 2020). The majority of commercial modules have efficiencies of 13–19 per cent with over a 25-year guaranteed lifetime and longer in practice;
- *thin films* (TFs) using amorphous Si (a-Si/c-Si); micromorph silicon multi-junctions (c-Si); cadmium-telluride (CdTe); and copper-indium-[gallium]-[di]selenide-[di]sulphide (CI[G]S). They have lower efficiencies of 6–12 per cent (with a target of 12–16 per cent by 2020).

In development:

- *concentrating PV* (CPV);
- *organic PV*;

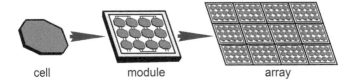

cell   module   array

Figure 4.1 The relationship between solar cells, modules and arrays

- *advanced thin films*;
- *multi-junction*.

Today's PV systems are fully competitive for off-grid electricity generation and with diesel-based on-grid systems in countries with good solar resources. In an increasing number of countries with high electricity costs and good solar resources, small PV systems are also achieving grid parity. c-Si systems accounted for 89 per cent of the market in 2011, the rest being TF.

There are many types of photovoltaic materials at various stages from research and development to fully commercial implementation. They each have different efficiency levels for converting solar energy into electricity. Other key characteristics are: operational lifetime, the effect of operating temperature, cost of production at scale and environmental impact of manufacture. Some features of the main types are listed in Table 4.1 and Figures 4.2 and 4.3.

4.2 Types of cell

4.2.1 *Wafer-based crystalline silicon technology*

Silicon is a chemical element with symbol Si and atomic number 14. It is the eighth most common element in the universe by mass. It occurs naturally in dusts, sands and rocks as various forms of silicon dioxide (silica) or silicates and over

Table 4.1 Performance of commercial PV technologies

	Cell efficiency (%)	Module efficiency (%)	Record commercial and (lab) efficiency (%)	Area/kW (m²/kW)	Lifetime (y) c-Si
c-Si					
Mono-c-Si	16–22	13–19	22 (24.7)	7	30
Multi-c-Si	14-18	11–15	20.3	8	30
TF					
a-Si	4–8		7.1 (10.4)	15	25
a-Si/_c-Si	7–9		10 (13.2)	12	25
CdTe	10–11		11.2 (16.5)	10	25
CI(G)S	7–12		12.1 (20.3)	10	25
Organic dyes	2–4		4 (6–12)	10 (15)	n/a
CPV	n/a	20–25	>40	n/a	n/a

Source: Data from EPIA, 2011

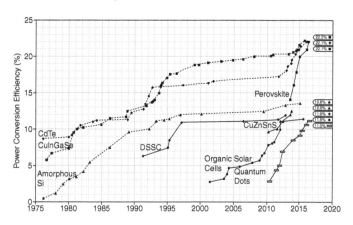

Figure 4.2 Best research cell efficiencies for the competing technologies

Source: www.ossila.com, with permission (2016)

90 per cent of the earth's crust is composed of silicate minerals. On its own it is a hard and brittle crystalline solid with a blue-grey metallic lustre. It is a tetravalent metalloid.

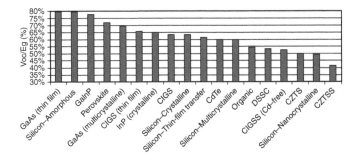

Figure 4.3 The maximum photon energy utilisation, defined as the open circuit voltage (Voc) divided by the optical bandgap (Eg), for common single-junction solar cells material systems

Source: www.ossila.com, with permission

Silicon is used in the three forms: single-crystal (sc-Si), multi-crystalline silicon (mc-Si) and ribbon-sheet grown c-Si. The majority of commercial modules are based on multi-crystalline silicon and low-cost manufacturing (screen-printing) and offer efficiencies between 12 per cent and 15 per cent (17 per cent in the best cases). Amorphous silicon (a-Si) does not contain crystals but atoms forming a continuous random network. The material is deposited in thin films onto a variety of flexible substrates, such as glass, metal and plastic. Amorphous silicon cells generally feature low efficiency.

Special processes for high-efficiency commercial cells include:

- buried contacts (by laser-cut grooves);
- back contacts (that currently achieve the highest commercial single-junction efficiency of 22 per cent); specialised surface texturing to improve sunlight absorption;
- HIT (heterojunction with an intrinsic thin layer), consisting of an sc-Si wafer between ultra-thin a-Si (amorphous silicon) layers to improve efficiency.

Table 4.2 Features of c-Si

c-Si material use	$3g/W_p$	
c-Si wafer thickness	<100mm	
Energy payback	0.5–1 years	
Cells per module	60–72	
Nominal power	120–300 W^p	
Surface area	1.4–1.7 m^2 (up to 2.5 m^2 maximum)	
Guarantee	sc-SI and mc-Si	80% of rated output for 20–5 years
	amorphous multi-junction	80% of rated output for 10–20 years
	amorphous thin film	Degrades 10% in first 3 months

Source: IEA

Single-junction solar cells have just one p-n junction, compared to multi-junction solar cells, which have more, made of different semiconductor materials, each of which will produce electric current in response to different wavelengths of light. This allows the absorbance of a broader range of wavelengths, improving efficiency.

Higher efficiencies have been achieved using materials other than silicon and multi-junction cells. The maximum theoretical efficiency for single-junction c-Si cells is currently estimated at around 29 per cent, but only a few commercial cells have efficiencies above 20 per cent. Module efficiencies are lower than those for cells. Current commercial single-junction sc-Si module efficiencies range between 13–19 per cent.

Power output

The power produced by a c-Si cell depends on its efficiency and the intensity of incident UV light. For comparison, modules are characterised by 'peak power' (watts-peak or Wp) output, as measured under standard test conditions (STC) – that is, the power produced when exposed to

Table 4.3 Surface area needed for different module types to generate $1kW_p$ under standard test conditions

Material	Surface area needed for $1kW_p$
Mono-crystalline silicon	$7-9m^2$
Polycrystalline silicon	$9-11m^2$
Amorphous silicon	$16-20m^2$

Source: IEA

1.0kW per square metre of sunlight at 25°C in an atmospheric air mass of 1.5.

Example: a module that is 10 × 10cm (0.1 m²) and 12 per cent efficient will produce 1.2 W under STC (10 per cent × 12 per cent × 1 kW). Therefore it would be rated at 1.2 Wp. If it were three times as big it would produce three times as much power: 3.6 Wp. If ten of them were connected together they would produce ten times as much power (12 Wp).

4.2.2 Thin film

Thin film photovoltaics (PV) have become a preferred alternative to the slower production and higher cost of traditional purified crystalline silicon wafers. They are called 'thin film' because the active layers are just a few microns (μm) thick, or about a tenth the diameter of a human hair. A thin layer of active materials is deposited on large area (m²-sized or long foils) substrates of materials such as steel, glass or plastic. They use small amounts of active materials and can be manufactured at a lower

Table 4.4 Features of thin film

Standard modules capacity:	60–120 Wp
Area:	0.6–1.0 m² (CIGS and CdTe)
	1.4–5.7 m² (silicon-based)

Source: IEA

overall cost than c-Si. They have shorter energy payback times (e.g. <1 year in southern Europe), good stability and lifetimes comparable to c-Si modules. Plastic TF are usually frameless and flexible and can adapt to different surfaces.

Multi-junction silicon (a-Si/m-Si) films offer higher efficiency than a-Si films. The basic material is combined with other active layers – for example, microcrystalline silicon (c-Si) and silicon-germanium (μc-SiGe), to form a-Si/μc-Si tandem cells, micromorph and hybrid cells (even triple junction cells) that absorb light in a wider range of frequency. An a-Si film with an additional 3m layer of μc-Si absorbs more light in red and near-infrared spectrum, and may reach efficiency up to 10 per cent. Commercial module efficiencies are <10 per cent.

Cadmium-telluride (CdTe) films offer efficiencies over 11 per cent. The world-record lab-based cell efficiency for a CdTe device is currently around 22 per cent. Solar panels based on CdTe are the only thin film photovoltaic technology to surpass crystalline silicon PV in cheapness for a significant portion of the PV market. Recent improvements have matched the efficiency of multi-crystalline silicon while keeping a cost advantage. The long-term availability of tellurium may be a concern.

Copper-indium-[gallium]-[di]selenide-[di]sulphide film (CI[G]S) offers better, over 13 per cent, efficiencies for commercial modules but they are costlier.

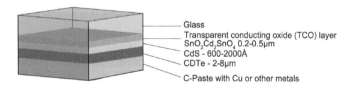

Figure 4.4 Cross-section through one design of CdTe cell

Source: Author, after NREL

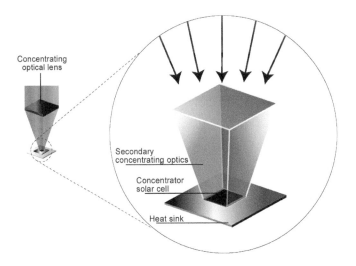

Figure 4.5 Schematic diagram of a CPV module utilising a lens

4.2.3 Concentrating solar power

Concentrating photovoltaics (CPV) are applicable at large scale between the 36° latitudes (see Figure 4.6), where clear skies predominate and normal direct irradiation is >1800 kWh/m²/year. Lenses concentrate sunlight from a wider area onto a smaller PV cell is located, making an array cheaper, or reducing the amount of land required, for the same amount of power output. Concentration can range from 4x–1500x. Systems include lenses, reflection and refraction systems. Efficiency depends on the degree of concentration and the area of the cell.

CPV can focus only the direct sunlight component, so high accuracy in optical and sun-tracking systems and heat dissipation are required. c-Si cells with efficiencies of 20–25 per cent are used with low–medium sunlight concentration while III-V semi-conductors and multi-junction solar cells (e.g. triple junction GaInP/GaInAs/Ge obtained from metal organic CVD) are used for high concentrations (>250).

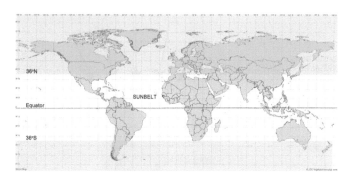

Figure 4.6 The Sunbelt area of the world, between the 36° latitudes. The darker the tint, the higher the insolation ($300–350W/m^2$)

Source: Author, from NASA data

4.2.4 *Organic PV*

These are based on the use of very low-cost materials and manufacturing processes, with low energy input and easy upscaling. These 'third generation' of electricity-producing solar cells will be market-ready by 2020. Three technologies are in use:

- oligomers, a molecule consisting of just a few monomers – molecules that may bind chemically or supramolecularly to other molecules to form larger ones;
- polymers, a large molecule, composed of many repeated subunits of oligomers;
- dye-sensitised solar cells (DSSC).

The ultra-thin layers of oligomer molecules are thermally evaporated onto the substrate, which is a flexible, semi-transparent plastic film. Ultra-light, less than 1mm thin, flexible and colourful, they can easily interface with organic LED technology – for example, for computer/phone displays. Oligomers are easier to manufacture than polymers, which require complex solvents and various printing processes.

On the cusp of a commercial availability, other advantages include no performance loss with rising temperatures; superior efficiency in low light levels, meaning they do not need to point at the sun; no toxins or heavy metals used in manufacture; easier to recycle; thin, flexible substrate; being transparent and colourful they have a multitude of potential applications, including the capability to be printed upon building cladding, glazing (windows), metals, membranes and gadgets. Current disadvantages include comparatively low efficiency, high cost and short life. Lab-based efficiencies for oligomers are 13 per cent. Encapsulation or some other filter is necessary to prevent oxidisation of the materials.

4.2.5 Advanced inorganic thin films

These include:

- evolutions of thin film concepts, such as glass beads covered by a thin multi-crystalline layer with a special interconnection between spheral cells and multi-crystalline silicon thin films obtained from high-temperature (>600°C) deposition;
- use of novel quantum effects and nano-materials enable a more favourable trade-off between output current and voltage of the solar cell;
- using up/down-converters to alter the irradiation frequency and maximise energy capture in existing solar cells. Photon absorption and re-emission shifts the sunlight wavelength and increases energy capture.

4.2.6 Perovskite

Perovskite solar cells (PSC), so called because of their crystal structure, hold great promise. Unlike traditional silicon cells, which require expensive, multistep processes conducted at high temperatures (>1000 °C) in a vacuum in special clean room facilities, the organic-inorganic perovskite material can be manufactured with simpler wet chemistry techniques in a traditional lab environment.

They have, at the time of writing, a certified conversion efficiency of 22.1 per cent.[1] The most commonly studied perovskite absorber is methyl-ammonium lead trihalide, which uses halogen atoms such as iodine, bromine or chlorine. Methylammonium and formamidinium lead trihalides have been created using a variety of solvent techniques and vapour deposition techniques, both of which have the potential to be scaled up. These techniques reduce the need to use so many polluting solvents.

Issues yet to be resolved are around stability, as the material can degrade within months, reducing its efficiency.

4.3 Power calculations

The maximum power obtainable from a solar cell is called the *maximum power point* (MPP) at which the voltage is known as V_m and the current I_m. The point is determined by intersection of the I–V curve of the solar cell and load line (which varies with temperature). For a constant resistive load the load line is a straight line with a slope I/R. The maximum power R_{opt} is obtained at point A, where:

$$R_{opt} = V_m / I_m$$

Temperature affects the efficiency of a c-Si cell as shown in Figure 4.8. Beyond 25°C (77°F), each degree Celsius of temperature rise causes a drop of around 0.5–0.6 per cent efficiency. At 38°C (100°F), a crystalline module will produce 6 per cent less power than the STC temperature of 25°C. The effect is less pronounced for thin film modules, for which it would be 2 per cent less.

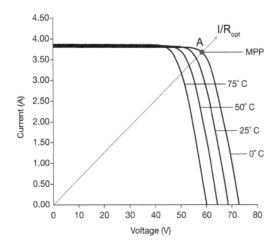

Figure 4.7 A solar module power curve, showing the maximum power point (MPP) and the effect of temperature on output

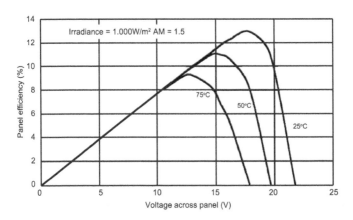

Figure 4.8 The performance of a photovoltaic cell at 25°C, 50°C and 75°C, receiving 1000W/m² in each case

Module manufacturers supply a *temperature coefficient* revealing how much power is lost per degree temperature increase. The lower the coefficient the more energy is produced in hot climates. Multiply the difference between the standard temperature (25°C) and the actual temperature by the coefficient and factor this into the power produced at the standard temperature to find the actual power produced. For example, a 300W-rated module with a −0.43%/°C temperature coefficient at 25°C and an irradiance of 1,000 W/m² generates 300W. If irradiance is constant but ambient temperature increases to 50°C, the same module now produces 300W × [1 + (50°C − 25°C) × −0.43%/°C] = 268W.

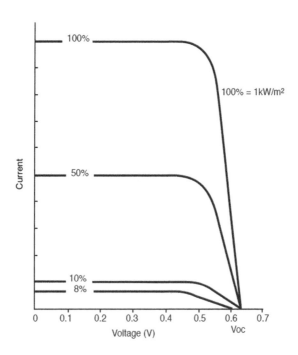

Figure 4.9 Variation in voltage and current by light intensity. Voltage is affected far less than current

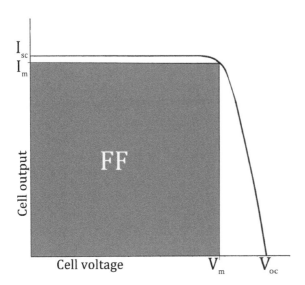

Figure 4.10 The fill factor

The *fill factor* (FF) is defined as the ratio of the maximum (peak power) from the cell to the product of the open circuit voltage (I_o) and short circuit current (I_{SC}), expressed as:

$$FF = \frac{V_m \times I_m}{V_{oc} \times I_{sc}} = \frac{P_m}{V_{oc} \times I_{sc}}$$ [57]

FF is a measure of the squareness of the I-V characteristic of the cell and the area of the largest rectangle that will fit inside the I-V curve (see Figure 4.10).

Conversion efficiency (η) is the ratio of the maximum power output (P_{opt}) to the input power (P_{in}) received at a given cell operating temperature.

$$\eta = \frac{P_{opt}}{P_{in}} = \frac{P_{opt}}{AG_T}$$ [58]

where A = cell area in m^2 and G_T = solar irradiance in W/m^2.

Cell output power across a load (P_L) is

$$P_L = I_L V = I_L^2 R_L \tag{59}$$

where I_L is the load circuit current, V is the voltage and R_L is the load resistance.

Open circuit junction voltage is:

$$V_{OC} = \frac{kT}{e_o} ln \left[\frac{I_{SC}}{I_o} + 1 \right] \tag{60}$$

Where:

k = Boltzman's constant (1.381×10^{-23} J/K)
T = cell temperature (K)
E_o = electron charge (1.602×10^{-18} J/V)
I_{sc} = short circuit current
I_o = reverse saturation current.

To calculate *available power* multiply the insolation (I) by peak power (W_p) by the performance ratio (r) of the system (η plus the efficiency of the whole system).

$$P = I.W_p.r \tag{61}$$

Example: if the insolation is 850kWh/m^2/annum (850 peak sun hours), the panels are rated at 120W_p and the performance ratio is 75 per cent, this gives 76,500Wh or 76.5kWh/annum. In practice, most locations do not receive a standard solar spectrum even in cloudless conditions – for example, due to dust etc.

4.4   Balance of system elements

The balance of the system (BoS) elements and costs are those other than the panels and include: installation, the inverter (to convert DC into AC), power control systems, cabling, racking, grid connection (including metering) and energy storage devices, if any.

Inverters are available with capacities from 100W to as much as 2 MW. Single or numerous inverters can be used for a single PV system. The main considerations are improved lifetime, matching to module requirements, reliability and control of reactive power in grid-connected systems. Choose one according to the peak load it will have to meet.

Charge controllers adjust the charge into batteries to ensure optimal performance. They include a low-voltage disconnect that prevents excessive discharging, which can damage the batteries. The best controllers include maximum power point tracking (MPPT), which optimises the PV array's output, increasing the energy it produces. The choice of controller depends on the size of the PV system and the system voltage.

Storage technologies include lithium-ion and valve-regulated lead-acid batteries; pumped hydro storage systems (large scale only); capacitors; hydrogen generation; compressed air systems.

- *For lead-acid batteries*, the 'deep discharge' type are preferred because they can be almost completely drained without too much damage. Avoid auto batteries.
- *Lithium-ion batteries* have the advantage of compactness, reliability and the ability to withstand more charge and discharge cycles but a higher upfront cost. Leasing arrangements are common. They dominate the market for portable devices and are attractive for electric vehicles and distributed renewable power, frequency regulation and UPS, with an energy density up to 630 Wh/l, cycle efficiency (90 per cent) and durability, and low self-discharge (5–8 per cent per month at 21°C).

Grid connection. Grid voltage and frequency regulation require a response time from seconds to minutes and higher power output. Daily load-levelling requires extensive power to be available for

Table 4.5 The performance of storage technologies

Storage type	Power (MW)	Discharge time	Efficiency (%)	Lifetime (years)	Overall storage cost (USD /MWh)	Capital cost (USD /kW)
Pumped hydro	250–1000	10h (depending on water storage volume)	70–80	>30	50–150	2,000–4,000 (100–300) b
Compressed air energy storage (CAES)	100–300 (10/20)	3–10h	45–60	30	–150	800–1,000 (1,300–1,800) c
Flywheels	0.1–10	15s–15m	>85	20	na	1,000–5,000 d
Supercapacitors	10	<30s	90	5×10^4 cycles	na	1,500–2,500 (500) d
Vanadium redox flow battery f	0.05–10	2–8h	75/80DC 60/70AC	5–15	250–300 d	3,000–4,000 (2000) d
Li-ion battery	~5	15m–4h	90DC	8–15	250–500 d,e	2,500–3,000 (<1,000) d,e
Lead battery	3–20	10s–4h	75/80DC 79/75AC	4–8	na	1,500–2,000
NaS battery	30–35	4h	80/85DC	15	50–150 d	100–2,000 d
Superconducting magnetic energy storage (SMES)	0.5+ d	1–100s/h d	>90	$>5 \times 10^4$ cycles	na	na

Note:
a. All figures are intended as typical order of magnitude estimated based on available sources and information, often with wide ranges of variability
b. Hydro power plant upgrading for storage service
c. Small systems (10–20 MW)
d. Projected/estimated
e. Large Li-ion cells
f. For grid use. Approaching commercial viability. Energy is stored in the electrolyte and safe and non-flammable. Modular and reliable

hours, and the cost it can bear depends on the marginal price of peak-load electricity. Investment costs for storage equipment are highly variable.

4.5 General PV system design advice

Approximate *PV system-sizing software* is available (see Chapter 11).
The following advice comes from experience of many systems:

- keep it simple: increased complexity reduces reliability and increases costs, especially for maintenance;
- appreciate that the system may not be available 100 per cent of the time. Achieving this makes the system more expensive: be realistic;
- for stand-alone systems, be realistic when estimating loads: including a large safety factor can increase costs substantially;
- repeatedly check weather data: errors in estimating the solar resource can cause disappointment;
- different hardware with different characteristics have different costs. They may also be less compatible. Investigate thoroughly all options before deciding on the optimum combination;
- ensure the system is installed carefully: each connection must be made to last 30 years, because it can if installed properly. Deploy the correct tools and techniques. System reliability is no higher than the weakest connection;
- electricity is dangerous: be rigorous about safety precautions during installation and in operation. Comply with local and national building and electrical codes;
- plan for periodic maintenance: PV systems have a good reputation for unsupervised operation but all require some degree of monitoring and care;
- calculate the lifecycle, levelised costs to compare PV systems with alternatives. This reflects the complete lifetime cost of owning and operating any energy system.

The modules should:

- not be shaded (even partly), especially between the peak sun hours of 10am to 3pm;
- have sufficient space to mount the required number, where there is no risk from damage or overheating, since efficiency tails off significantly at higher module temperatures;
- be accessible for regular cleaning;
- have a roof or mount able to withstand their weight, wind forces, plus the weight of any snowfall.

Note that failure of inverters is perhaps the most common fault.

4.6   Costs

Modules account for, on average, 50 per cent of total system costs (depending on application and technology). To reduce costs (£/kWh) it is necessary to:

- reduce the *balance of system* costs (system components and installation costs);
- increase the energy yields, stability and lifetime of the system;
- increase the inverter lifetime and reliability of system components;
- not combine modules of different specifications in the same system;
- match the inverters to the modules and load profiles.

The cost (or value) of solar electricity is much higher in the winter, when light is scarce, than in the summer. For example, 7.5 hours of sunlight might provide the same power, 20 Ah, in the summer, as 240 hours in the winter. In a grid-connected system when solar electricity is unavailable it is made up by purchasing grid power. In a stand-alone system it is made up by storage or back-up power.

Table 4.6 The range of capacity factors at utility scale

Plant type	Capacity factor range (%)
Natural gas combined cycle	40–64
Oil (steam turbine)	9–16
Hydroelectric	30–47
Coal	54–73
Nuclear	78–91
Wind farms	31–49
Geothermal	65–75
Photovoltaic solar (daytime)	28–33
CSP solar in California	33
CSP solar with storage and natural gas back-up in Spain	63
Landfill gas	65–72

Source: EIA

4.7   Capacity factor

The *capacity factor* of a power plant is defined as the ratio of the actual electricity produced per year to the electricity that could in theory be produced, based on the nominal peak power and technical availability of the PV system 24/7. It is given as a percentage.

4.8   Small-scale stand-alone system design

A typical, small stand-alone system will include the following:

- a photovoltaic module or array;
- a battery charge controller;
- batteries;
- safety disconnects and fuses;
- a grounding circuit;
- cables;
- an inverter or power control unit, if necessary to convert DC to AC;
- loads: e.g. fluorescent lamps rated from 6–11W and LEDs 3–5W.

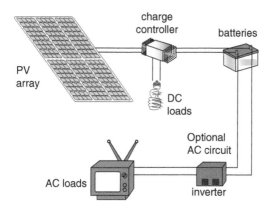

Figure 4.11 Schematic diagram of a basic stand-alone DC circuit with optional additional AC circuit. A single string of panels is connected in series. There will be a dedicated protective device inserted into any existing consumer unit. Array junction boxes (including string fuse disconnects) are situated between the arrays and the charge controllers. An array main fuse disconnect is situated between the charge controller and the batteries. An isolator is required either side of the inverter

Battery capacity in amp-hours (Ah) is calculated using:

$$Ah = \frac{Wh}{V_s} \cdot D \tag{62}$$

where Wh = the total load value in watt-hours, V_s = the voltage of the system and D = the number of days for which back-up is to be required. System design calculations are solved using software (see Chapter 11).

For a *grid-connected small system*, a PV generation meter is installed between the PV inverter and the main consumer unit. An additional, lockable AC isolator may be required by the district network operator between the PV generation meter and the main consumer unit.

4.9 Utility-scale PV

The US administration's National Renewable Energy Laboratory has set the definition of utility-scale solar PV at 5MW; the International Energy Agency (IEA) suggests 1MW, as does the US Energy Information Administration; authorities like the International Renewable Energy Agency use the term without defining it; since 2013 it has been set by WikiSolar at $4MW_{AC}$, roughly equivalent to $5MW_P$ (solar peak megawatts DC).

Solar parks where all areas receive identical insolation are often configured in discrete rectangular blocks, with a centralised inverter in a corner or the centre of the block. These may be of the order of 1MW each. Where insolation is not uniform across the site, or where the specification of the panels varies, string inverters are used. These are substantially lower in capacity, of the order of 10 kW. They condition the output of a single

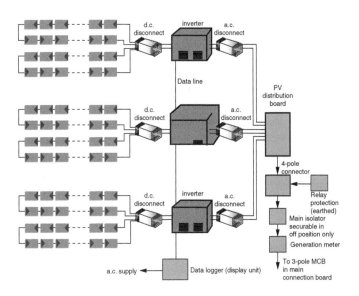

Figure 4.12 Schematic layout for a 3-phase modular utility-scale system using strings in parallel

array string – that is, a whole, or part of, a row of solar arrays within the overall plant.

4.9.1   Fixed or tracking arrays?

Tracking can be made in three ways:

- *vertical axis east–west only*: it will not rotate at a constant speed during the day;
- *horizontal axis only*: up–down;
- *two-axis*: both east–west direction and inclined, so that the modules always point at the sun.

For utility-scale projects a two-axis produces >20 per cent more energy than fixed-tilt, costs much less than 20 per cent more to build but takes up more space. On a given area of land it may be possible to fit 160 per cent more capacity using fixed-tilt, yielding more total energy. In this case calculating row spacing and tilt angle to optimise yield is done by spacing at the maximum permissible width for a 4 × 4 utility vehicle to gain access. Row-on-row shading can occur but typically only near dawn/dusk and in winter when irradiance and the price of electricity are low, so the value of lost energy is negligible. Lowering the tilt angle can allow for tighter row spacing without increasing row-on-row shading; lowering the height of the top module could lead to cost savings in mounting and maintenance.

4.9.2   Suggested time line for a project (with average time estimate)

PV plants are notable for the speed at which they can be designed, installed and commissioned. A reasonable average term is nine months. This suggested process is taken from a 1MW project. (Steps need not be sequential.)

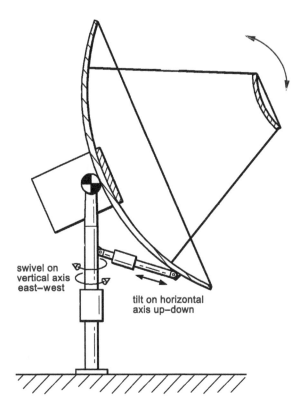

Figure 4.13 Tracking for a concentrating solar parabolic dish reflector up–
down and east–west. Stand-alone flat plate PV modules could
also track in the same manner. Parabolic troughs would normally only tilt on the horizontal axis in the up–down direction

Source: Author

1. Preparation of design and estimation of the plant, using site
   insolation data (15 working days).
2. Planning permission (12 months).
3. Land acquisition (1 month).

4. Power purchase agreement (PPA) with power purchaser, including technical requirements for grid connection (15 working days).
5. Preparation of detailed project report including technical feasibility (6 working days).
6. Arrangement of finance/fund with good interest rate, net metering and/or equity share (1 month).
7. Procurement work including preparation and finalising of vendor selection, BOM, BOQ, order placing, follow-ups of delivery to site (1 month).
8. Civil construction at the site for PV mounting structure set-up, control-room, administrative building (1 month).
9. PV installation and all electrical construction works, including the grid connection (1 month).
10. Installation of SCADA (Supervisory Control and Data Acquisition) system.
11. Commissioning of the plant by certified third party followed by completion of project (6 working days).

4.9.3   *Operation and maintenance*

A PV plant needs a regular maintenance schedule after commissioning. This includes regular cleaning of the arrays (depending on rain/dust levels), checking the status of inverters, cable fault checking, emergency maintenance and replacement of components. Self-cleaning coatings for panels are available. Financial planning therefore allows for the cost of the operation and maintenance (O&M) team. Some of this work may be automated – for example, fault checking.

4.10   Worked example

Modelling of a PV plant using two-axis tracked heliostats. This plant is located in Egypt.[2] To calculate the net power generation of the PV system, the solar resource and the incidence angle on the PV panel are needed. The measured global horizontal irradiation (G) and the sun elevation angle (θ) permit the calculation of the direct normal irradiation (DNI).

A PV system also converts diffuse irradiation so the calculated DNI values need to be converted into the global tilted irradiation value (GTI) depending on the solar angles at the particular time of the day. In order to simplify the calculation, the diffuse radiation on a horizontal plane D is normalised depending on the solar elevation angle (θ) to the measured diffuse normal irradiation DN and added to the DNI. This results in the GTI on the PV panel, assuming that the PV panels are arranged on a two-axis tracked heliostat:

$$DNI = \frac{G - D}{sin\theta}$$

$$DN = \frac{D}{sin\theta}$$

$$GTI = DNI + DN = \frac{G - D}{sin\theta} + \frac{D}{sin\theta} \qquad [63]$$

In order to account for the maximal tilt angle of the heliostats at sunrise and sunset, the sun elevation angles (θ) below 10° are set to zero. The resulting GTI values are significantly higher compared to the DNI values ranging up to 1,100 W/m². To check these figures they are compared with Meteonorm data and show a sufficient correlation.

To calculate the net power output of the complete PV plant, the first step is to calculate the MPP voltage (U_{mpp}) at given irradiation GTI. This is expressed in the following equation (all values with the subscript 0 describe the reference conditions given by the manufacturer in the PV panel's datasheet):

$$U_{mpp} = U_{mpp,0} \cdot \frac{1nGTI}{1nGTI_0} \qquad [64]$$

where the same applies for the MPP current (I_{mpp}) at irradiation level depending on the GTI:

$$I_{mpp} = I_{mpp,o} \cdot \frac{GTI}{GTI_o} \quad\quad\quad [65]$$

The PV panel heats up during operation. This reduces the power output. Usually, this effect is expressed by the heat-up coefficient (t_{PV}) which is given in K per solar irradiation using the unit $K/W/m^2$. The temperature of the PV panel (T_{PV}) consequently calculates to:

$$T_{pv} = T_{amb} + GTI \cdot t_{pv} \quad\quad\quad [66]$$

In order to calculate the MPP output (P_{mpp}) of the complete array, the number of PV panels switched in series (n_s) and in parallel (n_p) must be multiplied by the temperature correction term. The manufacturer gives a coefficient for the thermal properties of the PV panel ($\propto_{mpp}$) and usually given in %/K:

$$P_{mpp} = n_s \cdot n_p \cdot U_{mpp} \cdot I_{mpp} \left(1 + \alpha_{mpp} \cdot \left(T_{pv} - T_{amb}\right)\right) \quad\quad\quad [67]$$

The power output of the complete PV system is also influenced by three main losses which are given by efficiencies factors (η):

- the panel soiling (η_{soil}) respects the power reduction due to opacity;
- possible AC-DC inverter losses are respected by (η_{inv});
- and the total field efficiency is respected by (η_{fld}).

Adding the equations [63 to 66], the total power output of the PV system computes as follows:

$$P_{pv,net} = P_{mpp} \cdot \eta_{soil} \cdot \eta_{inv} \cdot \eta_{fld} \quad\quad\quad [68]$$

To specify the PV module first choose the module type to be purchased. This calculation assumes a crystalline silica-based module which has a long-term stable power output. A single module consists of 60 mono-crystalline solar cells mounted to a black anodised aluminium frame with a dimension of 1660 × 4990 mm. The operating temperature is given by the manufacturer in the range of −40 to +85°C. The soiling factor (η_{soil}) assumes a clean condition. An inverter has been chosen with 96 per cent efficiency. Table 4.7 summarises all parameters at reference conditions of GTI_o = 1000W/m^2 and $T_{amb,0}$= 25°C.

Now we know the characteristics of a single PV module and its power output we can calculate the total power output of the plant. In this particular example, the maximum design output capacity is decided at 2MW under nominal conditions. Using the values in Table 4.7 this computes to 8,800 modules. A single CSP heliostat fitted with mirrors has a reflective surface of 100m^2. Theoretically, given the module dimensions, replacing the mirrors with PV modules would allow for 60 PV modules on one heliostat. But to allow for the possible additional weight of the PV modules, just 55 PV modules per heliostat are decided upon. All the modules on a single heliostat are electrically connected in series. The heliostats themselves, each now

Table 4.7 Example PV module characteristics

Parameter	Symbol	Value	Unit
Type and model number	BOSCH c-Si M60S	M245	–
MPP output	P_{mpp}	245	W
MPP voltage	U_{mpp}	30.11	V
MPP current	I_{mpp}	8.14	A
Temperature coefficient	$\propto_{mpp}$	−0.44	%/K
Module area	$A_{pv,m}$	1.64	m^2
Heat-up gradient	t_{pv}	0.04	K/W/m^2
Soil factor	η_{soil}	96	%
Inverter efficiency	η_{inv}	96	%
Field efficiency	η_{fld}	98	%

Table 4.8 Example PV plant system parameters

Parameter	Symbol	Value	Unit
Heliostats with PV modules, in parallel	η_p	160	–
PV modules per heliostat, in series	η_s	55	–
MPP voltage, system	$U_{mpp,0,sys}$	1656	V
MPP current, system	$I_{mpp,0,sys}$	1302	A
MPP power, system	$P_{mpp,0,sys}$	2157	kW
PV system, total aperture area	$A_{pv,sys}$	14462	m^2
Annual power generation, total plant	$P_{tot,pv}$	5948	MWh
Annual power generation, per heliostat	$P_{hel,pv}$	37	MWh
Total number of heliostats (CSP and PV)	η_{tot}	1943	–

carrying 55 PV panels (n_p), are connected in parallel. The total array adds up to a nominal power generation of 2.16 MW at peak. The calculated totals of the complete plant are summarised in Table 4.8.

4.11   Safety issues

- PV modules produce electricity when exposed to daylight and individual modules cannot be switched off. Therefore, unlike most other electrical installation work, installation necessitates working on a live system.
- PV module string circuits cannot rely on normal fuse protection alone for automatic disconnection of supply under fault conditions, as the short circuit current is little more than the operating current. Once established, a fault may remain a hazard, perhaps undetected, for a considerable time.
- Proper and appropriate wiring design and installation will protect installers and O&M workers.
- Undetected fault currents can develop into a fire hazard. Proper d.c. system design and a careful installation is needed.
- Care must be taken due to the combination of hazards: risk of shock, work at height and manual work.
- Correct experience and training is essential. National standards must be followed. Safety labelling is vital.

- Load calculations must be made with regulated safety factors. Consult a structural engineer.
- Cable penetrations through a roof should not affect its weather-tightness.

Inspection and testing documentation for the a.c. side typically comprises three documents:

- electrical installation certificate;
- schedule of items inspected;
- schedule of test results.

Inspection and testing of the d.c. side of the PV system is done in accordance with the requirements of BS 7671 and BS EN 62446 *Grid connected photovoltaic systems*.

The verification sequence contained within BS EN 62446 includes:

- inspection schedule;
- continuity test of protective earthing and/or equipotential bonding conductors (if fitted);
- polarity test;
- string open circuit voltage test;
- string short circuit current test;
- functional tests;
- insulation resistance of the d.c. circuits.

The system user should be provided as a minimum with the information as described in BS EN 62446 *Grid connected photovoltaic systems: minimum requirements for system documentation, commissioning tests and inspection*. This is a summary of the information required:

- basic system information (parts used, rated power, installation dates etc.);
- system designer information;

- system installer information;
- wiring diagram, to include information on:
  - module type and quantities;
  - string configurations;
  - cable specifications – size and type;
  - over-current protective device specifications (where fitted) – type and ratings;
  - array junction box locations (where applicable);
  - d.c. isolator type, location and rating;
  - array over-current protective devices (where applicable) – type, location and rating;
  - details of all earth/bonding conductors – size and connection points;
  - details of any connections to an existing lightning protection system (LPS);
- details of any surge protection device installed (both on a.c. and d.c. lines), to include location, type and rating;
- AC isolator location, type and rating;
- AC over-current protective device location, type and rating;
- residual current device location, type and rating (where fitted);
- module datasheets;
- inverter datasheets;
- mounting system datasheet;
- operation and maintenance information, to include:
  - procedures for verifying correct system operation;
  - a checklist of what to do in case of a system failure;
  - emergency shutdown/isolation procedures;
  - maintenance and cleaning recommendations (if any);
  - considerations for any future building works related to the PV array (e.g. roof works);
- warranty documentation for PV modules and inverters – to include starting date of warranty; and
- period of warranty;
- documentation on any applicable workmanship or weather-tightness warranties;
- test results and commissioning data.

4.12   PV power for water pumping

The electric power needed from the PV system to be supplied to the input of the motor-pump unit, P_{EL}, is given by:

$$P_{EL} = \frac{P_H + P_T}{\eta_{MP}} \qquad [69]$$

where η_{MP} is the efficiency of the motor-pump unit, and P_H (hydraulic power required to pump the water in watts) + P_f (friction in the pipes) equates to the mechanical power at the output of the pump.

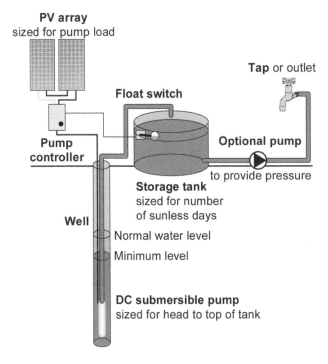

Figure 4.14 Schematic diagram of a PV-powered pump

P_H is given by:

$$P_H = g \cdot Q \cdot H_v \qquad [70]$$

where H_V is the apparent vertical head in (m), Q is the water flow rate expressed in (m^3h^{-1}) (numerically equal to mass-flow rate since the specific gravity of water is unity) and g is the acceleration due to gravity (2.725).

The electric power from a PV system (array plus inverter) is given by:

$$P_{EL} = P_{NOM} \cdot \frac{G}{G_{REF}} \cdot \eta_A \cdot \eta_I \qquad [71]$$

where P_{NOM} is the power of the PV array under standard test conditions (Irradiance = 1,000 W/m^2, AM 1.5, cell temperature = 25°C), G is the on plane irradiance, G_{REF} is the irradiance at STC, η_A is an array performance factor considering cell temperature, wiring and mismatch losses, and η_I is the inverter efficiency. The volume of water pumped throughout the day is given by:

$$Q_d = \int_{day} \frac{P_{NOM} \cdot G \cdot \eta_A \cdot \eta_{MPI}}{2.725 \cdot G_{REF} \cdot H_T} \, dt \qquad [72]$$

where $\eta_{MPI} = \eta_{MP} * \eta_I$. But this is a tough equation to solve. PV pump manufacturers usually provide standardised graphic tools relating water output with PV array power, under given radiation conditions and for constant pumping head. But daily irradiance may not be constant. A proven alternative (L. Narvarte, E. Lorenzo and E. Caamaño[3]) is to find an 'equivalent total head', H_{TE},

$$H_{TE} = H_{OT} + H_{ST} + \left(\frac{H_{DT} - H_{ST}}{Q_T} \right) \cdot Q_{AP} + H_F \left(Q_{AP} \right)$$

with $Q_{AP} = \alpha \cdot Q_d$ \qquad [73]

where $\alpha = 0.047$ (h^{-1}) when Qd is expressed in m^3 and $H_F(Q_{AP})$ is the head loss in the pipes corresponding to Q_{AP} – an average 'apparent flow rate'.

4.13   Social and environmental aspects of PV

4.13.1   Energy payback

Energy payback is the amount of time it takes for a panel to generate the same amount of energy that was expended in its manufacture. Note that much of the energy for manufacture is heat, which may be considerably cheaper than electricity, and that the figure depends greatly on particular types and manufacturers.

Factoring in the energy payback of the mount and system components would add, on average, another 15 per cent of time for c-Si and 30 per cent for thin film. These are average figures. In reality it will depend upon factors such as the location of the panels, the amount of insolation they receive, system efficiency etc.

4.13.2   Carbon payback

Carbon payback is the time (or amount of energy generated) it takes for a renewable energy product or system to offset the same amount of greenhouse gas emissions that were caused by its

Table 4.9 Energy payback by cell type for one type of manufacture

Cell type	Energy payback (years)	Total energy that is pollution-free (%)
c-Si – polycrystalline	2.2	91
c-Si – mono-crystalline	1.7	93
a-Si – ribbon	1.6	93
CdTe	0.7	97
DSSC	0.4	98

Source: James Durrant, Imperial College, London

manufacture. This depends upon the fuel mix of the local grid supplying the factories. In an area where electricity is predominantly generated by burning coal, carbon payback will take significantly longer than if the local grid is predominantly renewably sourced.

A 2009 survey examined the global warming impact of glass/glass-encapsulated thin film amorphous silicon-based modules and nano-structured materials in thin film silicon modules, used in roof-mounted modules installed in southern Europe, lasting 30 years (a comparatively rare type of module). It was found that the lifecycle impact of amorphous PV systems was between 0.03kg and 0.04kg CO_2/kWh and for micromorph thin film between 0.06kg and 0.08kg CO_2/kWh. This is considerably less than the impact of natural gas at 0.195kg CO_2/kWh and electricity from the (UK) national grid at 0.55kg CO_2/kWh (based on the annual average).[4]

4.13.3   *Environmental effects of photovoltaics*

Many toxic chemicals are used during the manufacture of PV cells. A report[5] by the Silicon Valley Toxics Coalition (SVTC) says that there may be no cause for alarm about their impact in countries with well-regulated environmental pollution regimes, but such strict monitoring and controls are not necessarily in force everywhere. Every year, the SVTC produces a scorecard of how well manufacturers are performing in making their products safer and less polluting. This scorecard may be used to help choose a manufacturer. The top three suppliers for 2014 were Trina, Sunpower and Yingli. The Solar Scorecard is based on SVTC's annual survey of photovoltaic (PV) module manufacturers, as well as on prior survey responses, interviews, news stories and publicly available data.

At final decommissioning, modules should be sent to specialist waste treatment; however, as yet, this provision is rare (see below) – final disposal of modules is most unusual, since displaced modules can be used in other locations.

Since 2010, 44.5 per cent of the PV industry (based on 2013 market share) has participated in one or more SVTC surveys. Seven companies representing 25.2 per cent of the PV module market share responded to the 2014 SVTC survey; 37 responded to the scorecard.

The problems identified include: greenhouse gas emissions; the need for workers in plants making the panels to be protected from toxic exposure; and preventing hazardous e-waste dumping in developing countries like India, Ghana and China, where the proper infrastructure to protect workers or the environment is lacking.

Silane gas, used in the production of c-Si, is extremely explosive and presents a potential danger to workers and communities. Accidental releases of the gas have been known to spontaneously explode; the semiconductor industry reports several silane incidents every year. The production of silane and trichlorosilane results in waste silicon tetrachloride ($SiCl4$), an extremely toxic substance that reacts violently with water, causes skin burns and is a respiratory, skin and eye irritant. This will only be a problem in places with little or no environmental regulation.

Sometimes the extremely potent greenhouse gas sulphur hexafluoride (SF_6) is used to clean the reactors used in silicon production. The Intergovernmental Panel of Climate Change (IPCC) considers sulphur hexafluoride to be the most potent greenhouse gas per molecule; one ton of sulphur hexafluoride has a greenhouse effect equivalent to that of 25,000 tons of CO_2. It is being phased out in most of the industry. Other chemicals used in the production of *crystalline silicon* requiring special handling and disposal procedures include:

- sodium or potassium hydroxide (used to remove the sawing damage on the silicon wafer surfaces): dangerous to the eyes, lungs and skin;
- hydrochloric acid, sulphuric acid, nitric acid and hydrogen fluoride (used to remove impurities from and clean semiconductor materials);

- phosphine (PH_3) or arsine (AsH_3) gas (used in the doping of the semiconductor material): inadequate containment or accidental release poses occupational risks;
- phosphorous oxychloride;
- phosphorous trichloride;
- boron bromide and boron trichloride.

Other dangerous chemicals used in the manufacture of *a-Si* cells include acetone, aluminium, chlorosilanes, diborane, phosphine, isopropanol, nitrogen, silicon tetrafluoride, tin and, where germane is used, germanium and germanium tetrafluoride.

CdTe cells use cadmium, cadmium sulphide, cadmium chloride and thiourea. Cadmium is a known carcinogen[46] and is considered 'extremely toxic' by the US Environmental Protection Agency (EPA) and Occupational Safety and Health Administration (OSHA). The Pesticide Action Network recognises thiourea as a 'bad actor chemical' because it is a known carcinogen and can be toxic.

Numerous chemicals are used in the production of *CIS* and *CIGS* panels, many of them very toxic. These include hydrogen selenide (or selenium hydride, H_2Se), which is considered highly toxic and dangerous at concentrations as low as 1 part per million in the air.

Extended producer responsibility (EPR) requires companies to take responsibility for the impacts of their products: from the materials used in manufacturing to product recycling. One well-known form of EPR is producer take-back, which requires companies to take back their products when users are done with them and ensure that they are recycled safely and responsibly. EPR policies provide incentives for companies to design and produce cleaner and more easily recyclable products, and discourage the practice of 'planned obsolescence' (intentionally making products that quickly become out of date or useless).

Whereas most PV modules sold in Europe are covered by a prefunded EPR scheme to ensure safe and responsible disposal, no PV modules in the USA are. Three PV manufacturers (Trina, Yingli and Up Solar) have written to the Solar Energy Industries Association (SEIA) seeking action on EPR for PV modules in the USA. Over the past three SVTC surveys, 14 companies have said they would support public policy for an EPR scheme for PV modules. The SVTC comment that 'commercial, government, or residential purchasers of PV modules are making a long-term financial and environmental commitment, and PV module manufacturers should make the same long-term commitment to the environment and worker safety'.[6] SVTC also seeks to stop the practice of sending e-waste to US prisons for dismantling, which results in toxic exposure to inmates.

The good news is that emerging technologies for the generation of solar electricity are not likely to be so potentially toxic. These include dye-sensitised solar cells, which utilise a pigment that effectively absorbs sunlight – the dye can be organic, of plant origin, like the colouring found in pokeberries or blackberries; when light falls onto the dye-sensitised solar cell, it is absorbed by the dye – and organic solar cells, made of biodegradable materials.

Notes

1 Woon Seok Yang et al, Iodide management in formamidinium-lead-halide–based perovskite layers for efficient solar cells, *Science* (July 2017). doi:10.1126/science.aan2301

2 Wellmann, J., and Morosuk, T., 'Renewable energy supply and demand for the City of El Gouna, Egypt', *Sustainability*, 8 (4), 314 (2016). doi:10.3390/su8040314

3 Narvarte, L., Lorenzo, E., and Caamaño,E., 'PV pumping analytical design and characteristics of boreholes', *Solar Energy*, 68 (1), 49–56 (2000).

4 Alsema, E., and van der Meulen, R., 'Fluoride gas emissions from amorphous and micromorphous silicon solar cell production:

emission estimates and LCA results', Copernicus Institute/Utrecht University, the Netherlands, 2009. Available at: opus. kobv.de/zlb/volltexte/2010/8346/pdf/3889.pdf, accessed October 2010.

5 *Towards a Just and Sustainable Solar Energy Industry* (PDF) available at http://bit.ly/1wJZEh0, accessed December 2014.

6 Ibid.

5 Solar thermal systems

5.1 Collector types: small/medium scale

Solar thermal collectors absorb the sun's heat and transmit it to a fluid which is drawn off to transport the heat to where it can do useful work. At the small and medium scale there are three types of collector:

- *flat plate*: fluid-carrying pipes, set upon a plate, both coated with a selective surface for high absorption, lie inside a glass-fronted box that is insulated at the sides and below. The glass is coated to maximise absorption on the outside surface and to reflect heat back inside on the inside surface;
- *polymer-based*: these are flat plate panels that are made of one moulded piece of plastic. These are cheaper and less efficient;
- *batch or integral collector storage system*: the collector and one or more tanks are combined in the same insulated glazed box situated on a roof. This is simple and cheap and used in warmer climes where freezing is rare;
- *evacuated tube*: an absorber, usually in the shape of a fin, with a selective coating is enclosed in a sealed glass vacuum tube. The tubes are arranged in rows, each connected to a manifold at the head where a heat exchanger transfers the heat to a fluid. The manifold conveys the fluid from all

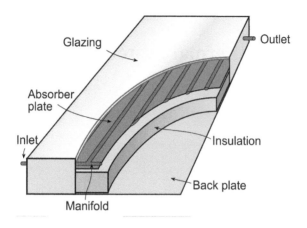

Figure 5.1 Flat plate collector

Source: Author

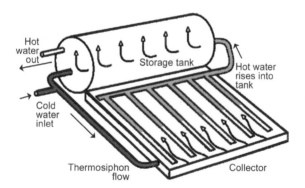

Figure 5.2 Batch system solar collector

Source: Author

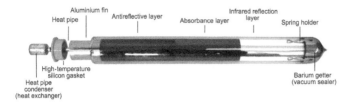

Figure 5.3 Single evacuated tube collector

Source: Author

Figure 5.4 Evacuated tube solar collectors on a roof at Panorama, Thessaloniki, Greece

Source: Green Solar, licensed under the Creative Commons Attribution-Share Alike 3.0

the tubes to the storage tank. These collectors are more expensive but more efficient and often more cost-effective, being ideal for delivering moderate to high temperatures. There are two main types of evacuated tube:

1. *direct flow*: the fluid is also circulated through the piping of the absorber;
2. *heat pipes*: the absorbed heat is transferred using the heat pipe principle.

For more information see section 5.6.4.

5.2   Small-scale solar thermal system types

At a small scale, the above collectors may be used in active or passive, and open or closed (direct or indirect) systems. Refer to Table 5.1.

System elements typically include:

- collector array support structure;
- hot water storage tank;
- a pump required to transfer the fluid to the tank (except in thermosyphon systems and outdoor swimming pools, which utilise the filtration system pump);
- valves, strainers and thermal expansion tank;
- controller to switch on pump only when useable heat is available (except in thermosyphon systems or if a photovoltaic-powered circulator is used);
- freeze protection low-toxic heat transfer fluid in the loop between the solar collector fluid and heat exchanger in the tank;
- safety features such as overheating protection, seasonal systems freeze protection or prevention against restart of a large system after a stagnation period.

Table 5.1 Summary of small solar thermal system types and their pros and cons

	Open/direct	Closed/indirect	Advantages	Disadvantages
Active (pumped)	The water heated is water used	The heat is transferred from a closed loop	The tank can go anywhere below the collector	Requires pump, electricity source, valve/thermostat controller
Passive (thermosyphoned, no pump)	The water heated is the water used	The heat is transferred from a closed loop	Does not require a pump; more reliable Used within a latitude of +/−20°; low maintenance; long lasting	The storage tank must be situated above top of collector(s) If cloudy cannot provide 24/7 hot water – needs supplemental heat source
Advantages	Easiest to install; works where it never freezes; works for swimming pools; low maintenance	Can be used where the temperature drops below freezing; water being heated indirectly can be in an insulated tank indoors and reach higher temperatures		
Disadvantages	Unsuitable where it freezes unless system is drained; cannot reach high temperatures; water heated is the water used	More complex; requires a heat exchanger, affecting system efficiency; heat-transmitting fluid (which contains antifreeze) must be non-toxic		

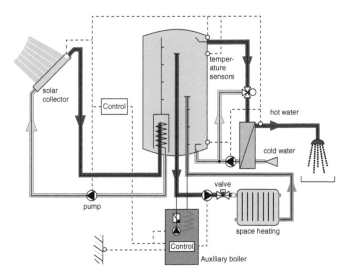

Figure 5.5 Schematic diagram of an active, closed loop system

Source: Author

5.2.1 Open loop system

For areas where freezing is rare. The controller switches on the pump when the collector water is hotter than the return from the tank.

5.2.2 Closed loop system

Used in non-tropical countries. There is a choice between two layouts: fully filled and drainback.

Fully filled

Includes an expansion vessel in the circuit for the collector fluid. A one-way check valve is positioned after the pump and before the expansion vessel and the solar collector.

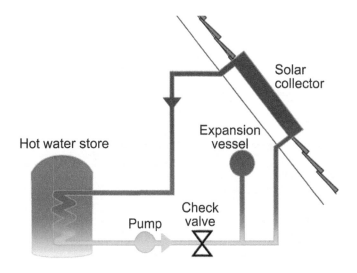

Figure 5.6 A fully filled, closed loop solar thermal system layout with an expansion vessel to contain the fluid, which expands when very hot

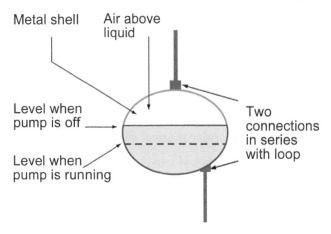

Figure 5.7 The behaviour of the expansion tank (which must not be insulated)

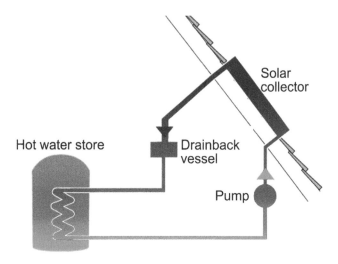

Figure 5.8   A drainback, closed loop solar thermal system layout. Air is
present in this closed circuit, so when not being pumped the
water falls back into the drainback vessel

Drainback

Expansion of the fluid is contained by a vessel with an air pocket.
Requires a more powerful pump than a fully filled system. Used if
there's a danger that fluid left in the collector when it is not
circulating might reach extremely high temperatures.

5.3   Medium-scale solar thermal system design

A medium-scale system would provide hot water for heating a
facility or commercial building or apartment block. It will be
integrated with another heating system. In such a system, solar
heat is used to heat water which can be topped up by other heat
sources if and when it is not sufficient. The design variables for
sizing the system can be summarised as:

Metal casing   Internal flexible
shell

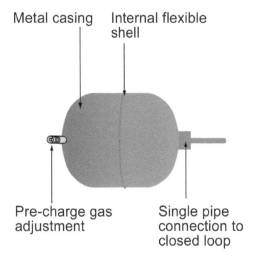

Pre-charge gas   Single pipe
adjustment   connection to
                           closed loop

Figure 5.9 The drainback vessel in a drainback solar thermal system.
    It is insulated

- hot water load requirement;
- integration with existing water heating or mechanical equipment;
- collector type, size, angle, racking and mounting;
- heat exchanger;
- flow rates, temperatures and pressures;
- storage capacity;
- pipe and pump sizes;
- fluid and pipe expansion;
- solar radiation;
- equipment layout;
- system location and altitude;
- aesthetics.

5.3.1   System efficiency

The *energy factor* (EF) is an indicator of system efficiency. It is the amount of hot water energy supplied divided by the amount of supplemental fossil-fuel energy used over a prescribed time period. A higher number indicates greater efficiency. It is calculated by:

$$EF = Q / E + E_o \qquad [74]$$

where:

Q = heat delivered by the water heating system
E = supplementary electricity (gas) energy used by the storage tank (burner) in addition to the solar energy
E_o = ancillary electrical energy to operate a circulating pump and controls, if used.

5.3.2   Tank sizing

Correct sizing of the system is critical. Too much heat generated can be very dangerous, yet system oversizing can lead to very high maintenance costs or complete failure. A system should be sized based on providing no more energy than needed to recharge the store of hot water in periods of low demand, normally in the summertime. Providing a larger store may cause low temperatures to occur when solar heat is scarce. While a greater solar fraction may be available at some times, this oversizing risks causing disappointment.

The reader is referred to the ASHRAE charts (at www.ashrae.org) to benchmark the typical usage pattern for their type of heating system. Metering is strongly advised: water meters, natural gas meters, BTU meters, ultrasonic flow meters and dataloggers may all be deployed.

Storage size is specified as a proportion of the collector surface area, the temperature required, the total load, the local climate and available solar irradiance. Ratios of collector to tank size are often specified in local building regulations. For example, in the USA it is 1 square foot of collector to 1.5–2 gallons of storage volume for both commercial and domestic systems. In England it is $1m^2$ to 25 litres.

5.3.3 *Pump sizing*

Pressure drops in systems should be minimised with thoughtful pipework design and correct pump sizing. Pressure drops can cause the fluid to boil and are dangerous.

The 'head' is the height above the suction inlet that a pump can lift a given volume of fluid. Pump head (H) can be converted to pressure (P) using the specific gravity (SG) of the fluid by the equation:

$$P = 0.434 \cdot H \cdot SG \qquad [75]$$

or by the density of the fluid (ρ) and the acceleration due to gravity (g):

$$P = H \cdot \rho \cdot g \qquad [76]$$

When selecting centrifugal pumps, the rated pump head must be equal to or greater than the total head of the system (total dynamic head or TDH) at the desired flow rate.

Net positive suction head available (NPSHA) is the physical measure of a fluid's ability to resist cavitation. Cavitation happens inside a pump when the local pressure falls below the vapour pressure of the liquid being pumped, causing the liquid to boil. This must be avoided. Comparing that value with the NPSHR (R = Required) makes it possible to specify a suitable pump. NPSHR is a function of the pump and is provided by the pump manufacturer. NPSHA is a function of the system and must be calculated. It can be calculated as:

$$NPSHA = APH + / - SPH - VPH - FHL \qquad [77]$$

where:

APH is atmospheric pressure head
SPH is suction pressure head
VPH is vapour pressure head
FHL is friction head loss in the suction piping.

Atmospheric pressure depends on elevation and temperature. An online calculator for revealing the atmospheric pressure, given these variables, can be found at http://bit.ly/2sLmCqm

Net head is proportional to the power delivered to the fluid, called output power (P_{out}), which describes the useful work the pump will do. It can be calculated by the equation:

$$P_{out} = \dot{m} \cdot g \cdot H = \rho \cdot g \cdot Q \cdot H \qquad [78]$$

where:

ρ is fluid density
g is the acceleration due to gravity
Q is the volumetric flow rate
H is the pump head
$\dot{m}$ is the mass flow rate.

In all pumps there are losses due to friction, internal leakage, flow separation etc. Therefore the power supplied to the pump, called the input power (P_{in}), is always larger than the water horsepower. This specification is typically provided by the manufacturer as a rating or in a performance curve. It is used to select the right motor or power source for the pump. The most energy efficient pump should be selected, ideally with a variable speed drive.

Pump efficiency (η_{pump}) defines the percentage of energy supplied to the pump that is converted into useful work.

It is the ratio between the useful power and the required power:

$$\eta_{pump} = P_{out} / P_{in} \qquad [79]$$

Keep in mind that any efficiency rating of the pump given by the manufacturer assumes system conditions which may not be identical to yours. Manufacturers usually designate an optimum or best efficiency point (BEP) of the performance curve. Plotting the system curve on this same graph will reveal an intersection point of the two curves, which is the operating point of the pump in the system. The ideal pump is one in which the required operating point intersects at the pump's BEP. Consider the use of solar-powered pumps.

5.3.4 *Safety issues*

System design and installation must address the following:

- prevention of Legionella bacteria developing within the consumed drinking or shower water. Auxiliary heating is used;
- scalding risk from steam or hot water: failsafe control should be installed to keep temperatures within safe limits at water outlets;
- prevention of stagnation of water, often a result of system oversizing;
- addition of glycol or other measure to prevent liquids freezing;
- potential water treatment to tackle possible accumulation of solids or bacteria, particularly in direct systems;
- installation of a check valve to prevent backflow or thermosiphoning of heated water into a cold water cistern;
- minimising the disturbance of stratification in the solar storage vessel during sterilisation cycles;
- loss of liquid through overflow. Any expansion vessel must be sufficiently large to prevent stagnation causing fluid loss;
- in closed systems, providing a heat dump in the form of a radiator to dissipate surplus heat.

5.4   Collector types: large scale

Large-scale collectors are deployed in Sunbelt regions – for example, North Africa, Middle East, south-western USA, southern Europe. They use systems which concentrate the sun's energy onto smaller areas to achieve higher temperatures. There are five variant technologies:

1. parabolic trough (PT);
2. Fresnel reflector (FR);
    both concentrate incident sunlight on a focal line. Maximum operating temperatures: 300–550°C;
3. solar tower (ST);
4. solar dish (SD);
    both focus on a single focal point and may reach higher temperatures;
5. solar updraft towers: in the theoretical stage.

5.4.1   Parabolic trough

Arrays of troughs focus heat onto a pipe filled with fluid (synthetic thermal oil, direct steam or molten salt). Commercial

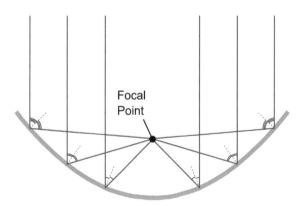

Figure 5.10 Schematic diagram of a parabolic trough reflector

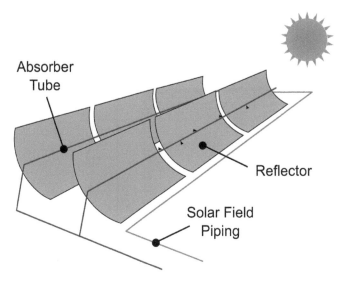

Figure 5.11 Schematic diagram of a parabolic trough collector field

PT plants exist with capacities of 14–80MWe with *maximum operating temperature* of 390°C; *efficiency*: 14–16 per cent; *capacity factor*: 25–30 per cent; *investment cost*: $4,200–8,500/kW dependent upon local conditions, DNI, maturity level of project and presence of thermal storage; *levelised cost*: $115–200/MWh with storage, $330/MWh without storage; *lifetime*: over 30 years. Over 90 per cent of installed global concentrating solar thermal power (CSP) capacity consists of PT plants.[1]

5.4.2 *Fresnel reflector*

Concentrating linear Fresnel reflectors (CLFRs) use series of ground-based, flat or slightly curved Fresnel mirrors placed at

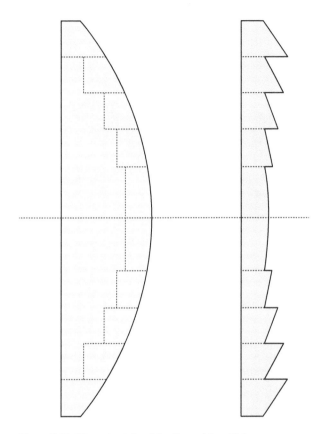

Figure 5.12 The principle of the Fresnel lens/dish

different angles to concentrate sunlight onto a fixed receiver located several metres above the mirror field. Each line of mirrors employs a single-axis tracking system. The receiver consists of a long, selectively coated tube where flowing water is converted into saturated steam – direct steam generation (DSG). The focal line can be distorted by astigmatism, so either a secondary mirror is placed above the receiver to refocus the sun's rays or multi-tube receivers can be used to capture sunlight with no secondary mirror. The advantage compared to PT is the lower cost of mirrors and solar collectors (including structural supports and assembly), but the optical efficiency is lower. Whether FR electricity is cheaper than that from a given PT would be determined through modelling and costing.

Solar Fire, an appropriate technology NGO in India, has developed an open source design for a manually operated, 12 kW peak Fresnel concentrator that generates temperatures up to 750°C (1,020K, 1,380°F) and can be used for various thermal applications including steam-powered electricity generation.

5.4.3 Solar tower

Solar tower (ST) plants have higher concentration factors and use water-steam (DSG), synthetic oil or molten salt as the primary heat transfer fluid. High-temperature gas is also a possibility. DSG in the receiver eliminates the need for a heat exchanger between the primary heat transfer fluid and the steam cycle, but makes thermal storage tougher.

Maximum operating temperatures: 250–300°C (using water-steam); 390°C (using synthetic oil); <565°C (using molten salt); >800°C (using gases). The temperature level of the primary heat transfer fluid determines the operating conditions (i.e. subcritical, supercritical or ultra-supercritical) of the steam cycle in the generator. *Thermal storage medium*: molten salt at 565°C.

High-temperature ST plants offer potential advantages over other CSP technologies in terms of efficiency, heat storage, performance, capacity factors and costs.

5.4.4 Solar dishes

A parabolic dish-shaped concentrator reflects sunlight into a receiver at the focal point. The receiver may be a Stirling engine (i.e. kinematic and free-piston variants) or a microturbine. Two-axis sun tracking systems are necessary. Very

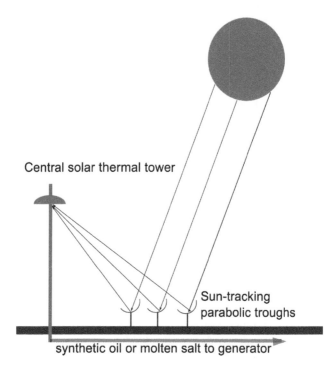

Central solar thermal tower

Sun-tracking
parabolic troughs

synthetic oil or molten salt to generator

Figure 5.13 Schematic diagram of a solar tower system

high concentration factors and operating temperatures are possible. *Advantages*: high efficiency (up to 30 per cent); modularity (in 5–50 kW units); suitable for distributed generation; no need for cooling systems for exhaust heat, so suitable for water-constrained regions. *Disadvantages*: relatively high electricity generation costs compared to other CSP options; still under demonstration; investment costs high. Existing plants have capacities from 10–100 kW. One, the Big Dish in Australia, uses an ammonia-based thermo-chemical storage system.

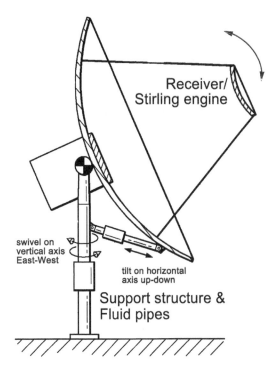

Figure 5.14 Schematic diagram of a solar dish collector

5.4.5 *Solar updraft tower*

This uses a selective absorbance cover to heat the air beneath it in an enclosed cavity that is shaped to direct the hot air by convection to a central tower. The higher the tower the more uplift is generated and the greater the speed of the air. The air drives a conventional wind turbine to generate electricity. Currently experimental and likely to be of relatively poor efficiency compared to the above CSP technologies.

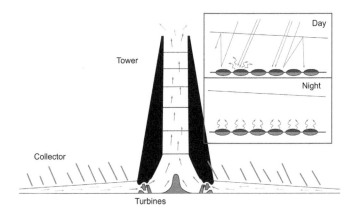

Figure 5.15 Schematic diagram of a solar updraft tower

5.5 Large-scale solar system types

Commercial-scale solar thermal plants heat water to drive steam turbines to generate electricity. Synthetic oil, steam or molten

Table 5.2 Summary of all solar thermal collector types

Type	Concentration ratio	Indicative operating temperature (°C)
Unglazed flat plate absorber	1	40
Flat plate collector	1	60–120
Fixed concentrator	3–5	100–150
Evacuated tube	1	5–180
Compound parabolic	1–5	70–180
(with one axis tracking)	5–15	70–290
Parabolic trough	10–50	150–350
Fresnel refractor	10–40	70–270
Spherical dish reflector	100–300	70–730
Parabolic dish reflector	200–500	250–700
Central receiver	500–3,000	500–1,000

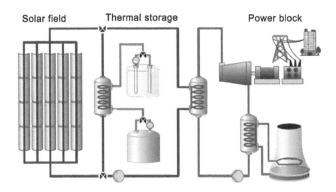

Figure 5.16 Schematic diagram of a CSP plant with storage. Excess heat is sent to the heat exchanger and warms the molten salts going from the cold tank (top) to the hot tank. When needed, the heat from the hot tank can be returned to the heat transfer fluid and sent to the steam generator to produce electricity or heat or both

salt are used to transfer superheated steam to power a Rankine cycle generator.

In arid, coastal regions CSP can also be used for water desalination by the production of electricity for reverse-osmosis water desalination or heat for thermal distillation.

The annual full load hours that can be supplied by CSP vary based on the level of thermal storage, latitude and annual solar irradiation (DNI). See Table 5.3.

Plants can be equipped with thermal storage to generate electricity after dark or under cloudy skies. The storage medium can be water or molten salt (more common); capacity can be up to 15 hours.

To provide sufficient thermal power to heat the storage medium as well as provide power during the day, the solar field (i.e. mirrors and heat collectors) must be oversized with respect to the nominal electric capacity (MW) of the plant. This is called the oversize factor or solar multiple. Given maximum

Table 5.3 Annual full load hours of a CSP plant for different solar multiple, latitude and level of annual direct normal irradiance (DNI in kWh per m² per year) (h per year)

Latitude	DNI 1800	DNI 2000	DNI 2200	DNI 2400	DNI 2600	DNI 2800
6,000 m² per MW, no storage:						
0°	1,613	1,869	2,128	2,362	2,594	2,835
10°	1,607	1,859	2,130	2,344	2,581	2,808
20°	1,559	1,801	2,082	2,269	2,502	2,725
30°	1,460	1,689	1,977	2,128	2,350	2,580
40°	1,310	1,524	1,815	1,920	2,127	2,366
12,000 m² per MW, 6-h storage:						
0°	3,425	3,855	4,221	4,645	4,931	5,285
10°	3,401	3,817	4,187	4,612	4,909	5,222
20°	3,310	3,719	4,098	4,495	4,810	5,096
30°	3,147	3,539	3,943	4,283	4,605	4,887
40°	2,911	3,285	3,719	3,984	4,301	4,604
18,000 m² per MW, 12-h storage:						
0°	4,869	5,414	5,810	6,405	6,713	7,147
10°	4,829	5,358	5,752	6,365	6,690	7,074
20°	4,711	5,223	5,630	6,229	6,583	6,929
30°	4,499	4,995	5,434	5,970	6,352	6,676
40°	4,189	4,674	5,163	5,601	5,987	6,322
24,000 m² per MW, 18-h storage:						
0°	5,987	6,520	6,796	7,563	7,859	8,243
10°	5,918	6,430	6,711	7,514	7,831	8,160
20°	5,761	6,260	6,563	7,380	7,724	8,009
30°	5,506	5,999	6,340	7,110	7,497	7,738
40°	5,155	5,650	6,045	6,717	7,115	7,348

Source: Trieb and others 2009[2]

solar irradiance (*c*.1kW/m²) the oversize factor or solar multiple should be 1.1–1.5 (up to 2.0 for linear Fresnel reflectors), depending mostly on the amount of sunlight the plant receives and the sun's variation throughout the day. Plants with large storage capacities may have solar multiples of 3–5.

Operation and maintenance (O&M) costs of CSP plants are low compared to those of fossil fuel-fired power plants. A typical 50 MW parabolic trough plant requires about 30–40 employees for operation, maintenance and solar field cleaning. Automation can reduce the O&M costs, including fixed and variable costs, and insurance, by more than 30 per cent.

5.6 Evaluating collectors

The remainder of this chapter is concerned with technical calculation and may be skipped if software is being used to size systems.

5.6.1 Unglazed collectors

Unglazed collectors are described by the following equation:[3]

$$\dot{Q}_{coll} = (F_R\alpha)[G + (\frac{\varepsilon}{\alpha})L] - (F_R U_L)\Delta T \qquad [80]$$

where:

$\dot{Q}_{coll}$ is the energy collected per unit of collector area per unit of time

ε is the longwave emissivity of the absorber

L is the relative longwave sky irradiance defined as:

$$L = L_{sky} - \sigma(T_\alpha + 273.2K)^4 \qquad [81]$$

where:

L_{sky} is the longwave sky irradiance

T_a the ambient temperature expressed in degrees Celsius

$F_{R\alpha}$ and $F_R U_L$ are a function of the wind speed V incident upon the collector. The values of $F_{R\alpha}$ and $F_R U_L$, as well as their wind dependency, are specified by the user.

The wind speed incident upon the collector is set to 20 per cent of the free stream air velocity specified by the user. The ratio ε/α

is set to 0.96. Because of the scarcity of performance measurements for unglazed collectors, a 'generic' unglazed collector is also defined as:

$$F_R \alpha = 0.85 - 0.04V$$
$$F_R U_L = 11.56 + 4.37V$$
[82]

5.6.2   Flat collectors

Duffie and Beckman's (1991[4], eq. 6.17.2) formula is one way to evaluate all types of non-concentrating collectors:

$$\dot{Q}_{coll} = F_R (\tau \alpha) G - (F_R U_L) \Delta T$$
[83]

where:

$\dot{Q}_{coll}$ is the energy collected per unit collector area per unit time
F_R is the collector's heat removal factor
τ is the transmittance of the cover
α is the shortwave absorptivity of the absorber
G is the global incident solar radiation on the collector
U_L is the overall heat loss coefficient of the collector and
ΔT is the temperature differential between the working fluid entering the collectors and outside.

Values of $F_R (\tau \alpha)$ and $F_R U_L$ are specified by the user and are independent of wind for both glazed and evacuated collectors. 'Generic' values can be used for glazed and evacuated collectors. For glazed collectors these are $F_R (\tau \alpha) = 0.68$ and $F_R U_L = 4.90$ (W/m²)/°C. For evacuated collectors these are $F_R (\tau \alpha) = 0.58$ and $F_R U_L = 0.7$ (W/m²)/°C.

G might need to be modified to account for the incidence angle. This is done substituting:

$$\tau \alpha = K_{\tau \alpha} \cdot (\tau \alpha)_n$$
[84]

where $K_{\tau\alpha}$ is the incidence angle modifier.

The European Standard EN12975 holds that the influence on $\tau\alpha_n$ by $K_{\tau\alpha}$ on the efficiency is:

$$\eta = F'K_\theta(\tau\alpha)_{en} - a_1 \frac{t_m - t_a}{G} - a_2 G\left(\frac{t_m - t_a}{G}\right)^2 \qquad [85]$$

So the relationship between $K_{\tau\alpha}$ and the efficiency is:

$$K_{\tau\alpha} = \frac{\eta(\theta)_{(at\ t_m - t_a = 0)}}{F'(\tau\alpha)_{en}} \qquad [86]$$

For conventional flat plate collectors, only one angle of incidence is needed in these definitions, which is 50°.

5.6.3 Concentrating collectors

The efficiency of a concentrating collector η can be deduced by considering both thermal radiation properties and Carnot's principle. For a solar receiver providing a heat source at temperature T_H and a heat sink at temperature $T°$ (e.g. atmosphere at $T° = 300$ K):

$$\eta = \left(1 - \frac{\sigma T_H^4}{C}\right) \cdot \left(1 - \frac{T^0}{T_H}\right) \qquad [87]$$

C = the concentration ratio.

To find the optimum temperature T_{opt} for which the efficiency is maximum:

$$T_{opt}^5 - (0.75T°)\,T_{opt}^4 - \frac{T°C}{4\sigma} = 0 \qquad [88]$$

Solving this equation numerically will obtain the optimum process temperature according to the solar concentration ratio

C – the ratio between the aperture area and the receiver area. It is necessary to measure the incidence angle effects from more than one direction (longitudinal and transversal) and at different angles (e.g. 20°, 40°, 60° and others) to fully characterise the incidence angle modifier. It can be estimated by considering it to be the product of the longitudinal and transversal incident angle modifiers, $K_{\theta L}$ and $K_{\theta T}$.

5.6.4   Evacuated tubes

These may be used to provide: domestic hot water; space heating and cooling (to drive a chiller); and process heating applications – typically at 60°C to 80°C, depending on outside temperature. Evacuated tubes can work with combi boilers to provide hot water on demand but require a storage tank between the tubes and the boiler. The tank output preheats the boiler, then feeds into it. If a sensor detects the water in the tank is already at the required temperature the output is directed straight to where it is required, bypassing the boiler.

How to calculate the heat output: conductive heat transfer between two surfaces having low-pressure gas (almost a vacuum) in the interim space is given by:[5]

$$q_l = k\Delta t \, / \, (g + 2p) \qquad\qquad [89]$$

where q_l is the heat loss, k is the constant, Δt is the temperature gradient, g is the gap between surfaces and p is the mean free path of molecules.

For air, the mean free path at atmospheric temperature and pressure is about 70 μm. If 99 per cent of air is removed from a collector, the mean free path increases at a substantially greater rate than the heat transfer path length (gap between the glass tubes) – c.20mm – reducing the conductive heat transfer substantially. The relative reduction in heat transfer as a function of the mean free path can be derived from:

$$\frac{q_{vac}}{q_l} = \frac{1}{1 + 2p / g}$$ [90]

where q_l is the conductive heat transfer if convection is suppressed and q_{vac} is the conductive heat transfer under vacuum. The effective heat gain of the collector based on the aperture area can be expressed as:[6]

$$q_u = (\tau\alpha)G_{eff} \frac{A_{tb}}{A_{cl}} - U_L(T_{abs} - T_a)\frac{A_{abs}}{A_{cl}}$$ [91]

where q_u is the useful heat gain (W/m^2) and G_{eff} is the effective solar radiation, both intercepted directly and after reflection from the back reflector (reflected radiation is typically $10\%(W/m^2)$); A_{tb} is the projected tube area (m^2), A_{cl} is the total collector area (m^2), U_L is the overall heat loss coefficient (W/m^2K), T_{abs} is the absorber temperature (°C), T_a is the ambient temperature (°C) and A_{abs} is the projected area of the absorber (m^2). Bekey and Mather have shown that a tube spacing of one diameter apart maximises the energy output because it allows sufficient solar radiation to heat the tubes.[7]

It is necessary to measure the incidence angle effects from more than one direction (longitudinal and transversal) and at different angles (e.g. 20°, 40°, 60° and others) to fully characterise the incidence angle modifier. It can be estimated by considering it to be the product of the longitudinal and transversal incident angle modifiers, $K_{\theta L}$ and $K_{\theta T}$.

5.7 Solar thermal system calculations

5.7.1 Calculating thermal energy requirements

The amount of energy needed to heat a volume of liquid from one temperature level to another is calculated in an identical manner to calculating the amount of thermal energy stored in any given tank. Both can be calculated as:

$$E = C_p \cdot d_t \cdot \rho \cdot V \qquad\qquad [92]$$

where:

E = energy (kJ, Btu)

c_p = specific heat capacity (of water = *4.19 kJ/kg°C Btu/lb °F*)

d_t = temperature difference between water stored and the surroundings (°C, °F)

ρ = mass (density) of liquid (kg, lb_m)

V = the volume of liquid.

By definition, it takes 1 kCal to raise 1kg (~1 litre) of water by 1 degree C (1,000 kCal = 1.163 KWh). The density of water changes according to temperature as shown in Table 5.4. *E* is allocated *pro rata* according to the number of days the system is used per week.

The performance of a solar water heating system with a storage tank depends on system characteristics, available solar radiation,

Table 5.4 Density of liquid water at 1 atm pressure

Temp (°C)	Density (kg/m³)
−30	983.854
−20	993.547
−10	998.117
0	999.8395
4	999.9720
10	999.7026
15	999.1026
20	998.2071
22	997.7735
25	997.0479
30	995.6502
40	992.2
60	983.2
80	971.8
100	958.4

ambient air temperature and heating load characteristics. The RETScreen Software Solar Water Heating Model can be used worldwide to evaluate the energy production and savings, costs, emission reductions, financial viability and risk: https://www.nrcan.gc.ca/energy/software-tools/7465

5.7.2 Calculating flow rates in heating systems

Besides temperature, the volumetric flow rate in a heating system is the other parameter it is necessary to know in order to regulate a heating system and achieve the required heating output. It can be expressed as:

$$q = \frac{h}{c_p \cdot \rho \cdot d_t} \qquad [93]$$

where:

q = volumetric flow rate
h = heat flow rate, defined as the energy (E, from above) required per unit of time (e.g. kwH) and the other variables are as above.

The next step is to determine how much energy can be provided by solar power.

5.7.3 The f-chart method

The f-chart method was originally developed as a way to predict the performance of solar thermal systems for small-scale applications at a time when the processing power needed for hour-by-hour simulations was unavailable. Its methodology could be completed using a scientific calculator. It is based on empirical correlations of the results of thousands of hour-by-hour simulations. The methodology is now available as software from several sources on the internet: for example, the program written by S.A. Klein and W.A. Beckman, the originators of the f-chart method. What follows is a description of the method.

The proportion of the space heating (if used) and hot water demand provided by a solar heating system (*solar fraction*) is known by the symbol *f*. It is a function of two dimensionless parameters, X and Y, where X is the ratio of collector losses to heating loads and Y is the ratio of absorbed solar radiation to the heating loads:

$$X = \frac{\text{Reference collector energy loss during a month}}{\text{Total heating load during a month}}$$

$$Y = \frac{\text{Total energy absorbed on the collector plate during a month}}{\text{Total heating load during a month}}$$

[94]

The *f-chart* method was developed with a standard storage capacity of 75 litres of stored water per square metre of collector area. For other storage capacities X has to be multiplied by a correction factor X_c/X defined by:

$$\frac{X_c}{X} = \left[\frac{actual\ storage\ capacity}{standard\ storage\ capacity} \right]^{-0.25}$$

[95]

It is arrived at as follows. First, X and Y are defined:

$$X = \frac{A_c F'_R U_L (T_{ref} - T_a)}{L}$$

$$Y = \frac{A_c F'_R \left(\overline{\tau\alpha}\right) H_T N}{L}$$

[96]

where:

A_c is the collector area
F'_R is the modified collector heat removal factor
U_L is the collector overall loss coefficient
T_{ref} is an empirical reference temperature equal to 100°C

T_a is the monthly average ambient temperature

L is the monthly total heating load

$\left(\overline{\tau\alpha}\right)$ is the collector's monthly average transmittance-absorptance product

H_T is the monthly average daily radiation incident on the collector surface per unit area and N is the number of days in the month;

F'_R accounts for the effectiveness of the collector storage heat exchanger.

The ratio F'_R/F_R is a function of heat exchanger effectiveness ε.

$$\frac{F_{R'}}{F_R} = \left[1 + \left(\frac{A_c F_R U_L}{\left(\dot{m}C_p\right)_c} \right) \left(\frac{\left(\dot{m}C_p\right)_c}{\varepsilon\left(\dot{m}C_p\right)_{min}} - 1 \right) \right]^{-1} \qquad [97]$$

where: $\dot{m}$ is the flow rate and C_p is the specific heat. Subscripts c and min represent the temperature of the collector-side and minimum of collector-side and tank-side of the heat exchanger. If there is no heat exchanger, F'_R is equal to F_R. If there is a heat exchanger, the assumption is that the flow rates on both sides of the heat exchanger are the same. The specific heat of water is 4.2 (kJ/kg)/°C, and that of glycol is set to 3.85 (kJ/kg)/°C. It is assumed that the ratio $A_c/\dot{m}$ is equal to 140 m² s/kg; this value is computed from thermodynamics collector test data (area 2.97m², test flow rate 0.0214 kg/s – Chandrashekar and Thevenard, 1995[8]). This equation is valid for ratios of actual to standard storage capacities between 0.5 and 4.

Finally, to account for the fluctuation of supply water temperature T_m and for the minimum acceptable hot water temperature T_w, X has to be multiplied by a correction factor X_{cc}/X defined by:

$$\frac{X_{cc}}{X} = \frac{11.6 + 1.18T_w + 3.86T_m - 2.32T_a}{100 - T_a} \qquad [98]$$

where T_a is the monthly mean ambient temperature. The monthly solar fraction f can then be determined from the following relation (obtained by regression from a detailed computer simulation – Okafor and Akubue, 2012[9]) as a function of X and Y:

$$f = 1.029Y - 0.065X - 0.245Y^2 + 0.0018X^2 + 0.0215Y^3 \quad [99]$$

If the formula predicts a value of f less than 0, a value of 0 per cent is used; if f is greater than 1 (100 per cent), a value of 1 is used.

The f-chart method proceeds as follows:

1. the variables X and Y are calculated for each calendar month;
2. the intersection of the X and Y values in Figure 5.17 locates a value of f which is the fraction of the heating load supplied by solar energy for each month. The fraction of the annual heating load supplied by solar energy f is the sum of the product of monthly solar energy fraction f and the monthly thermal load L_i divided by the annual load L.

$$f = \frac{\Sigma f_i L_i}{\Sigma L_i} \quad [100]$$

5.7.4   Piping and tank losses

For systems like heated pools without storage, the energy delivered by the solar collector, Q_{dld}, is equal to the energy collected Q_{act} minus piping losses, expressed as a fraction f_{los} of energy collected (f_{los} is entered by the user):

$$Q_{dld} = Q_{act}(1 - f_{los}) \quad [101]$$

For systems with storage, the load L (here termed Q_{load}) used in the *f-chart* method is made to include piping and tank losses to find $Q_{load,tot}$ as follows:

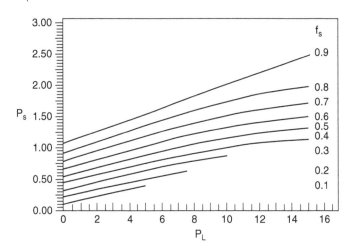

Figure 5.17 f-chart for liquid-based solar heating systems. f is the solar fraction, P_S is the absorbed solar energy divided by the heating load, and P_L is the reference collector loss divided by the heating load.

$$Q_{load,tot} = Q_{load}(1 + f_{los})$$
[102]

5.7.5 The utilisability method

This method will determine whether a system is able to supply sufficient energy from the available solar irradiance to overcome system losses. The monthly average daily utilisability $\bar{\phi}$ is dependent upon the collector area and orientation and the monthly radiation data. For a glazed collector it is found from the following four equations:

$$G \geq \frac{F_R U_L (T_i - T_a)}{F_R(\tau\alpha)}$$
[103]

where T_i is the temperature of the fluid entering the collector; G is the hourly irradiance in the plane of the collector; and all other variables have the same meaning as above. This makes it possible to find the monthly irradiance level G_c which must be exceeded for the system to collect solar energy. Average monthly transmittance-absorptance $(\overline{\tau\alpha})$ figures for the location and average daytime monthly temperatures T_a are used.

$$G_c = \frac{F_R U_L (T_i - \overline{T}_a)}{F_R (\overline{\tau\alpha})} \qquad [104]$$

This yields the average daily energy Q collected during a given month:

$$Q = \frac{1}{N} \sum_{days} \sum_{hours} A_c F_R (\overline{\tau\alpha}) (G - G_c)^+ \qquad [105]$$

where N is the number of days in the month and the + superscript denotes that only positive values of the quantity between brackets are considered. The monthly average daily utilisability $\overline{\phi}$ is defined as the sum for a month, over all hours, of the radiation incident upon the collector that is above the critical level, divided by the monthly radiation:

$$\overline{\phi} = \frac{\sum_{days} \sum_{hours} (G - G_c)^+}{\overline{H_T} N} \qquad [106]$$

where H_T is the monthly average daily irradiance in the plane of the collector. Substituting this definition leads to the monthly useful energy gain Q being found from:

$$Q = A_c F_R (\overline{\tau\alpha}) \overline{H}_T \overline{\phi} \qquad [107]$$

5.8   Thermal storage systems

Heat storage attached to a solar thermal plant allows it to continue providing energy to a generator after the sun has set, allowing it to compete with fossil fuel plants. There are three types of storage media:

1.   *Sensible heat storage*: capacities from 10–50 kWh/t; efficiencies 50–90 per cent, dependent upon the specific heat of the storage medium and insulation. Costs: €0.1–10/kWh.
2.   *Phase change materials* (PCMs): higher storage capacity and efficiencies – 75–90 per cent. Usually based on a solid–liquid phase change with energy densities on the order of 100 kWh/m³. Costs: €10–50/kWh.

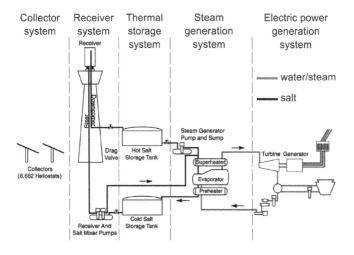

Figure 5.18 Schematic diagram of a high-temperature solar thermal system with storage

3. *Thermo-chemical storage* (TCS): capacities up to 250 kWh/t, dependent upon size, application and insulation. Costs: €8–100/kilowatt hour. Economic viability dependent partly upon the number and frequency of storage cycles and the material used.

5.8.1   Sensible heat storage

Sensible heat storage is the most commonly used method in solar heating systems. Storage media must possess high heat storage density, chemical stability and an ability to withstand repeated heating and cooling cycles. Water is most commonly used in low- and medium- temperature systems. For district heating systems, large underground tanks or caverns can be used. The earth, rock and concrete itself can be used as an insulating material. Heat storage can be calculated as:

$$\begin{aligned} q &= V \cdot \rho \cdot C_p \cdot dt \\ &= m \cdot c_p \cdot dt \end{aligned}$$

[108]

where:

q = sensible heat stored in the material (J, Btu)
V = volume of substance (m^3, ft^3)
ρ = density of substance (kg/m^3, lb/ft^3)
m = mass of substance (kg, lb)
c_p = specific heat capacity of the substance (J/kg°C, Btu/lb°F)
dt = temperature change (°C, °F).

5.8.2   Phase change materials (PCMs)

PCMs are used to store heat generated by concentrating solar thermal collectors, and for temperature moderation in buildings during hot weather. They yield or absorb more latent heat when

Table 5.5 Heat storage properties of different materials

Material	Temperature range (°C)	Density (ρ) (kg/m^3)	Specific heat (c_p) ($J/kg°C$)	Energy density ($kJ/m^{3o}C$)
Aluminium	max. 660 (melting point)	2,700	920	2,484
Brick		1,969	921	1,813
Cast iron	max. 1,150 (melting point)	7,200	540	3,889
Concrete	20	2,305	920	2,122
Fireclay		2,100–600	1,000	2,100–600
50% ethylene glycol – 50% water	0–100	1,075	3,480	3,741
Dowtherm A	12–260	867	2,200	1,907
Draw salt – 50% $NaNO_3$ – 50% KNO_3 (by weight)	220–540	1,733	1,550	2,686
Granite		2,400	790	1,896
Liquid sodium	100–760	750	1,260	945
Molten salt – 50% KNO_3 – 40% $NaNO_2$ – 7% $NaNO_3$ (by weight)	142–540	1,680	1,560	2,620
Therminol® 66	−3–343	750	2,100	1,575
Water	0–100	1,000	4,190	4,190

they change phase state. Oil or molten salt are used at high temperatures. The heat stored (Q) depends on the amount of storage material reacted (a_r), the endothermic heat of reaction (Δh_r) and the mass (m):

$$Q = a_r m \Delta h_r \qquad [109]$$

The storage capacity of a latent heat system with a PCM medium is given by:

$$Q = m\left[C_{sp}\left(T_m - T_i\right) + a_m \Delta h_m + C_{lp}\left(T_i - T_m\right)\right] \qquad [110]$$

(Sharma, Tyagi *et al.*, 2007[10])

where:

C_{sp} = average specific heat between the initial temperature T_i and the melting point T_m (kJ/kg K)

C_{lp} = average specific heat between T_m and the final temperature T_f (J/kg K)

a_m = the amount melted and

Δh_m = the heat of fusion per unit mass (J/kg).

For space cooling, a common PCM is paraffin-impregnated plasterboard. Micronal SmartBoard (30 per cent, 23°C), for example, is 5mm thick, containing 30 per cent (3 kg pr. m^2) microencapsulated PCM (paraffin) with a specific heat capacity of 1.20 kJ/kgK, a thermal conductivity of 0.18 W/mK (in solid state) and a latent heat capacity in the transition area of 330 kJ/m^2. The heat capacity for a volume of a material 'i' at time-step 'j' using a PCM like this is based on the temperature of the space in time-step 'j-1' – that is, the last known temperature. The heat capacity as a temperature change absorbed over time is calculated by:

$$(\rho c_p \Delta x_1)\frac{T_1^{j+1} - T_1^j}{\Delta t} = q_{surfside1} + \frac{T_{air} - T_1^{j+1}}{R_{surfside1}} + \frac{T_2^{j+1} - T_1^{j+1}}{\dfrac{\Delta x_1}{2\lambda_1} + \dfrac{\Delta x_2}{2\lambda_2} + R_2} \qquad [111]$$

(Rose *et al.*, 2009[11])

where:

Q = heat flux (W/m^2)

Δx = material volume width (m)

λ = thermal conductivity (W/mK)
R = thermal resistance (m²K/W)
H = specific enthalpy (J/kg)
Δt = size of the time-step (s)
$q_{surfside1}$ = heat transfer directly to the surface (W/m²)
$R_{surfside1}$ = thermal resistance for surface (m²K/W)
ρ = density (kg/m³)
c_p = heat capacity (J/kgK)

In CSP systems the use of molten salt for either storage or heat transfer reduces storage volume by up to 60 per cent, costs by 30 per cent and complexity compared to synthetic oil for heat transfer and molten salt for heat storage together. However, molten salt solidifies below 230°C and a heating system is needed during start-up and off-normal operation. Tanks of molten salt of around 29,000 tonnes each are used.

Table 5.6 Physical properties of some other phase change materials (PCMs)

Substance	Melting point (°C)	Conductivity (W/mK)	Density (kg/m³)	Specific heat (J/kgK)	Latent heat (kJ/kg)
Ice	0		0.92		333
Paraffin	−5–120		0.77		150–240
Hexadecane	20	0.39	777	1,390	281
Heptadecane	21	0.33	773		230
Dodecanol	24	0.28	853	1,550	235
Sodium sulphate	32.38		1,464		252
Erytritol	118		1.3		340

Sources: Yoon-Bok Seong and Jae-Han-Lim, 'Energy saving potentials of phase change materials applied to lightweight building envelopes', *Energies*, 6 (10) 5219–230 (2013), and Thermal Energy Storage, IEA-ETSAP and IRENA Technology Brief E17, January 2013, 9

5.8.3   *Thermo-chemical storage*

Table 5.7 Some interesting chemical reactions for thermal energy storage

Reaction		Temperature °C	Energy density (kJ/Kg)
Methane steam reforming	$CH_4+H_2O=CO+3H_2$	480–1,195	6,053
Ammonia dissociation	$2NH_3=N_2+3H_2$	400–500	3,940
Thermal dehydrogenation of metal hydrides	$MgH_2=Mg+H_2$	250–500	3079 heat storage; 9000 H_2 storage
Dehydration of metal hydroxides	$Ca(OH)_2=CaO+H_2O$	402–572	1,415
Catalytic dissociation	$SO_3=SO_2+{}^1/_2O_2$	520–960	1,235

Source: IRENA

5.8.4   *Storage options for larger systems*

- Lithium-based molten salts with high operation temperatures and lower freezing points.
- Concrete or refractory materials at 400–500°C with modular storage capacity and low cost (USD 40/kWh).
- Phase change systems based on Na- or K-nitrates to be used in combination with DSG.
- Cheaper storage tanks (e.g. single thermocline tanks), with reduced (30 per cent) volume and cost in comparison with the current two-tank systems.

Table 5.8 Comparison of features of solar thermal storage plants/storage media[12]

Medium	Molten salt	Concrete	PCM	Water/steam	Hot water
Capacity range (MWh)	500->3,000	1->3,000	1.0->3,000	1->200	1->3,000
Realised max. capacity of single unit (MWh)	1,000	2	0.7	50	1,000
Realised max. capacity of single unit (full load hours)	7.7	No data	No data	1	No data yet
Annual efficiency (%)	98	98	98	90	98
Heat transfer fluid	Molten salt	Synthetic oil, water, steam	Water/steam	Water/steam	Water
Temperature range (°C)	290.0–390.0	200–500	<350.0	<550	50–95
Investment cost (€/kWh)	40–60	30–40 (20 projected)	40–50 projected	180	2–5

Notes

1 Kraemer, S., 'Why CSP resurged in Africa and the MENA region', *Renewable Energy World*, 29 March (2016), see: http://bit.ly/2sp8kwn; also Solar Thermal Electricity Global Outlook 2016, Greenpeace International, European Solar Thermal Electricity Association (ESTELA), and SolarPACES.

2 Trieb, F., *et al.*, 'Global potential of concentrating solar power', presented at SolarPaces Conference Berlin, September (2009), available at: http://bit.ly/2sLWv2u

3 Soltau, H., 'Testing the thermal performance of uncovered solar collectors', *Solar Energy*, 49 (4), 263–72 (1992).

4 Duffie, J.A., and Beckman, W.A., *Solar Engineering of Thermal Processes*, 4th edn, Wiley, New Jersey (2013).

5 Goswami, D.G., *Principles of Solar Engineering*, 3rd edn, CRC Press, Boca Raton, CA (2015).

6 Michalsky, J., Stoffel, T., and Vignola, F., *Solar and Infrared Radiation Measurements*, CRC Press, Boca Raton, CA (2012).

7 Goswami (2015).

8 Chandrashekar, M., and Thevenard, D., Comparison of WATSUN 13.1 Simulations with Solar Domestic Hot Water System Test Data from ORTECH/NSTF – Revised Report, Watsun Simulation Laboratory, University of Waterloo (1995).

9 Okafor, I., and Akubue, G., 'F-Chart Method for Designing Solar Thermal Water Heating Systems', *International Journal of Scientific and Engineering Research*, 3 (9) (2012).

10 Sharma, A., Tyagi, V.V., Chen, C.R., and Buddhi, D., 'Review on thermal energy storage with phase change materials and applications' (2007), *Renewable and Sustainable Energy Reviews*, 13, 318–45 (2009).

11 Rose, J., *et al.*, 'Numerical method for calculating latent heat storage in constructions containing phase change material', paper presented at 11th International IBPSA Conference, Glasgow (2009).

12 Fichtner and DLR, *MENA Regional Water Outlook, Part II: Desalination Using Renewable Energy*, Fichter and DLR, Stuttgart (2011).

6 Desalination and drying

6.1 The need for desalination

According to the United Nations, by 2025, an estimated 1.8 billion people will live in areas plagued by water scarcity. Desalination of seawater is one solution to this crisis. It will become increasingly vital as a way of solving the world's acute water shortages.

But the conventional desalination process is energy-intensive. It emits, worldwide, around 76 million tons of carbon dioxide per year. Fossil fuel-powered power stations themselves use and waste much water. Saudi Arabia, for instance, uses around 300,000 barrels of oil every day to desalinate seawater, supplying 60 per cent of its fresh water this way.

A solution is to use renewable energy. The Global Clean Water Desalination Alliance aims to supply at least 10 per cent of the annual energy demand of existing water desalination plants with newly installed clean energy sources, including solar, by 2030. Solar-powered desalination is a highly sustainable choice, especially since much of the demand is in areas with high irradiance levels. Photovoltaic (PV) solar additionally has the lowest water consumption requirements.

The salt content of seawater is approximately 35g per litre (35,000 ppm) and it can reach concentrations of 39g per litre in certain seas in the world. To make fresh water, desalination techniques must reduce the salt concentration in the water to less than 1g per litre (1,000 parts per million or ppm).

6.2 Technologies for desalination

Three main techniques are presently used: reverse osmosis (RO), multi-flash distillation (MFS) and multiple effect distillation (MED). Further techniques are: mechanical vapour compression (MVC) and electrodialysis reversal (EDR).

Reverse osmosis: filters remove larger particles from seawater which is then pushed through semipermeable polymeric membranes at high pressure (54 to 80 bars) and ambient temperature. The membranes let through water molecules, but not larger particles such as dissolved salts, viruses and organic molecules. This produces about 50 per cent fresh water and 50 per cent brackish water. The salt content of the water is

Table 6.1 The energy required per litre of water produced by different desalination methods, both electrical and thermal, converting this to a total representing the electrical use as if it were all electrical

Desalination method	Electrical energy (kWh/m^3)	Thermal energy (kWh/m^3)	Electrical equivalent of thermal energy (kWh/m^3)	Total equivalent electrical energy (kWh/m^3)
Reverse osmosis	3–5.5	0	0	3–5.5
Multi-stage flash	4–6	50–110	9.5–19.5	13.5–25.5
Multi-effect distillation	1.5–2.5	60–110	5–8.5	6.5–11
Mechanical vapour compression	7–12	0	0	7–12

Source: 'Energy requirements of desalination processes', *Encyclopaedia of Desalination and Water Resources* (Desware) at www.desware.net

approximately 500 ppm. Much electricity is required to operate the pumps and the technique requires a high level of maintenance.

Multi-stage flash distillation: water is heated under pressure to about 120°C before being introduced to a series of low-pressure chambers which vaporise it. This requires electric power for pumps and much thermal energy to heat the seawater. There is no scaling. The salt content of the fresh water output is below 10 ppm.

Multi-effect distillation: water is heated to 70–80°C and introduced to a hot surface where it evaporates. Scaling can occur. Multiple effect evaporators with horizontal tubes are currently the most used devices. In these devices, the heating fluid flows in the horizontal tubes while seawater flows as a uniform film on the outside the tubes. The water evaporates and fresh water is then extracted from the tubes, while releasing its heat through condensation. A portion of the heating energy is thus recovered. The salt content of the produced fresh water is below 10 ppm.

Electrodialysis reversal: electricity is applied to electrodes to pull dissolved salts through an ion exchange membrane leaving fresh, low salinity water behind. This and high salinity concentrate are the products. The polarity of the electrodes is switched at fixed intervals to reduce the formation of scale and subsequent fouling and allow the system to achieve higher levels of fresh water recovery. A module contains an electrodialysis stack consisting of alternating layers of cationic and anionic ion exchange membranes. This system is less sensitive than RO to particulates and metal oxides and can remove arsenic, fluoride, radium and nitrates. Units also have a long membrane life (typically 20+ years for potable water installations).

Reverse osmosis accounts for 60 per cent of the global capacity for desalination, followed by multi-stage flash (MSF) (26.8 per cent). Desalination requires much energy, taking up around 50 per cent of total costs. The economics depend on the cost of the energy source.

6.3 Solar options for desalination

Solar power is increasingly being used to power these processes. The principal solar options are: concentrated solar power (CSP), solar thermal (ST) and photovoltaics (PV). Wind power is also an option (suffering from intermittency, however). Other solar options are currently in the experimental or pilot stages.

6.3.1 Concentrated solar power (CSP)

According to a recent World Bank report:[1]

> CSP is a competitive energy supply option… [and] the only economically viable renewable energy technology to store and provide power on demand. CSP is especially suitable to power desalination plants, most of which are required to operate around the clock.

CSP is an umbrella term covering four technologies: parabolic trough reflectors, solar power towers, parabolic dish collectors and linear Fresnel reflectors. See Chapter 5 for more details.

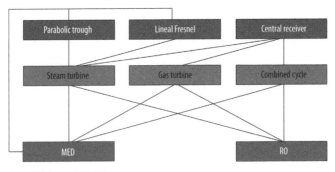

Source: Fichtner and DLR (2011)

Figure 6.1 Linking the choice of solar collection system to power generation and desalination[2]

Table 6.2 Comparison of concentrating solar power collecting systems[3]

Technology	Parabolic trough system	Linear Fresnel system	Solar power tower
Application	Superheated steam for grid-connected power plants	Saturated and superheated steam for process heat and for grid-connected power plants	Saturated and superheated steam for grid-connected power plants
Capacity range (MW)	10–250	5–250	10–100
Realized max. capacity single unit (MW)	80	2 (30 under construction)	20
Capacity installed (MW)	920 (1,600 under construction)	7 (40 under construction)	38 (17 under construction)
Peak solar efficiency (%)	21	15	<20
Annual solar efficiency (%)	10–16 (18 projected)	8–12 (15 projected)	10–16 (25 projected)
Heat transfer fluid	Synthetic oil, water/steam demonstrated	Water/steam	Air, molten salt, water/steam
Temperature (°C)	350–415 (550 projected)	270–450 (550 projected)	250–565
Concentration ratio	50–90	35–170	600–1,000
Operation mode	Solar or hybrid	Solar or hybrid	Solar or hybrid
Land use factor	0.25–0.35	0.6–0.8	0.2–0.25
Land use (m²/MWh/year)	6–8	4–6	8–12
Estimated investment costs (€/kW)	3,500–6,500	2,500–4,500	4,000–6,000
Development status	Commercially proven	Recently commercial	Recently commercial

	Molten salt, concrete, phase change material	Concrete for preheating and superheating, phase change material for evaporation	Molten salt, concrete, ceramics, phase change material
Storage options			
Reliability Advantages	Long-term proven • Long-term proven reliability and durability • Storage options for oil-cooled trough available	Recently demonstrated • Simple structure and easy field construction • Tolerance for slight slopes • Direct steam generation proven	Recently demonstrated • High temperature allows high efficiency of power cycle • Tolerates nonflat sites • Storage technologies are available, but still not proven in long term
Disadvantages	• Limited temperature of heat transfer fluid hampering efficiency and effectiveness • Complex structure, high precision required during field construction • Requires flat land area	• Storage for direct steam generation (phase change material) in very early stage	• High maintenance and equipment costs

Source: Fichtner and DLR 2011.[4]

Note: This comparison does not consider storage. If storage is considered, the central receiver applications with storage have the higher annual conversion efficiencies.

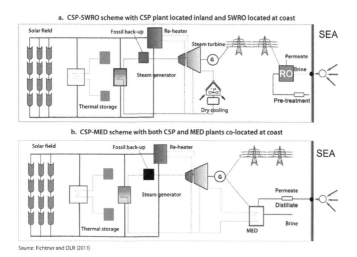

Source: Fichtner and DLR (2011)

Figure 6.2 Schematic diagrams for CSP and MED desalination plants[5]

Parabolic trough. Sun-tracking parabolic trough-shaped mirrors focus the sun's rays onto a pipe filled with water or mineral oil that is then pumped into a storage tank. This water is used to evaporate the water in a MED system. Some of the stored water is saved for the night-time so that the plant can run through the night. CSP can provide sufficient electricity to drive a conventional RO process, but MED is considered to be more suitable for integration in CSP plants since the exhaust steam from the turbine may be used to drive the MED process, increasing overall efficiency. Such has been the conclusion of techno-economic analyses in Abu Dhabi and Almeria, Spain. A single 400kW module can generate up to 65,000 gallons of fresh water per day in southern California. Seawater, wastewater, drainage water, run-off, saline groundwater or industrial process water may be treated this way. The remaining brine is concentrated into solid byproducts for resale.

Solar towers. Heliostats focus sunlight onto a central receiving tower. This contains a boiler which turns the water into superheated steam. This heat is used to evaporate the water in a MED system. Salt (as a phase change material – see Chapter 5) extracted from MED-desalinated water can be melted by the heat and used to store thermal energy. At night-time the heat can be drawn off, re-solidifying the salt, to continue powering the plant. Combining power generation and water desalination can be a cost-effective option for electricity storage when generation exceeds demand.

A pilot project in Cyprus is investigating co-generation of electricity and desalinated seawater from a solar power tower system. A large-scale commercial version of solar power tower technology that provides heating, fresh water and electricity is in place near Perth, Australia. It produces desalinated water for a greenhouse that grows about 15 per cent of Australia's tomato supply. The plant hosts a 234-tonne central boiler sitting on top of a 127 metre high tower that receives the reflected sun rays from over 23,712 mirrors (heliostats) of area 51,505m^2 to produce 36MWth of energy. Large plants like this can process >800,000 m^3/day. This technology can power MED or RO desalination.

Parabolic dish reflectors. These use similar technology but have a Stirling engine as a receiver at the focal point. They are more expensive but modular and cannot produce sufficient heat for thermal energy storage. They are not presently used for desalination.

With sufficient heat storage capacity, CSP is scalable to demand; it can provide both peak and baseload electricity; and with heat storage and oversized solar collectors, it can provide a firm power supply 24 hours a day. The efficiency of today's solar collectors ranges from 8–16 per cent, but, by 2050, technical improvements are expected to increase efficiency to the 15–25 per cent range. Currently, the solar energy collector field comprises more than half of the investment cost of a desalination system. Improvements in collection efficiency indicate significant potential for cost reduction.

The costs of fresh water produced by CSP thermal and RO membrane desalination plants also depend upon the seawater salinity level. In the Mediterranean and Red Sea regions, costs ranged in 2012 from US$1.52–1.74 per m³. Costs also vary depending on coastal or inland locations. Inland, higher solar radiation may reduce costs by as much as US$0.15 per m³.[6]

The World Bank puts the cost at c.US$1.8/m³, but says that by 2050 this is projected to fall to US$0.9/m³ (1 litre = 0.001 cubic metre).[7] It can already compete in regions where energy transmission and distribution costs are higher than those of distributed generation.

For comparison, the energy consumption of existing fresh water supplies including transportation over large distances is about 3 kWh/m³ but local fresh water supplies use 0.2 kWh/m³ or less. Large arrays of CSP might economically produce thermal and electrical power sufficient to desalinate over 5,000m³ water/day with energy storage. To minimise water use, hybrid cooling using a 25 per cent wet cooling tower and 100 per cent capacity dry cooling tower might be employed, where some turbine steam exhaust is reverted to the wet cooling tower when the ambient temperature rises.

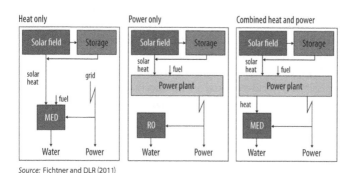

Source: Fichtner and DLR (2011)

Figure 6.3 Typical configurations of CSP desalination[8]

A bench study has estimated that if a CSP plant using MED desalination technology were constructed for desalination purposes, then the land area requirements for the collector array are $100km^2$ land area for 1billion m^3/y of water desalinating, or $1m^2$ collector area per $10m^3$ of fresh water derived through desalination.[9] This clearly depends on how much irradiation the area receives annually and how much thermal energy storage is required.

6.3.2   PV

PV costs have been falling, making PV a viable option in some circumstances to replace fossil fuel-generated electricity for powering a desalination plant. The first examples are coming on-stream. At Mohammed bin Rashid Al Maktoum Solar Park, near Dubai, a desalination facility already uses PV panels and batteries to produce about 13,200 gallons of drinking water a day, for use on site.

Much larger plants are in planning or construction phases. The world's first large-scale solar-powered desalination plant is being built at the time of writing in Al Khafji, Saudi Arabia. It will produce nearly 16 million gallons or 60,000 cubic metres of fresh water a day, enough to supply the local population of 150,000. Spanish solar company Abengoa and state-owned Saudi company Advanced Water Technology aim to complete it in 2017. To avoid the use of expensive batteries, it will generate a surplus of electricity during the day and draw a lesser amount from the grid overnight in order to run 24/7.

PV-powered desalination is currently up to three times as expensive as grid-powered plants, according to a World Bank report,[10] but this is partly dependent on the cost of land and battery storage. If land is cheap or existing roof space is used for the panels, and the grid used for storage, this improves the economics. Furthermore, PV systems are falling in cost and projected to drop from $50 per 1,000 gallons today in the Middle East to half of that by 2050. Government subsidies or carbon pricing can improve this picture.

A 10 megawatt advanced thin film PV farm exists south of Perth, in Western Australia. Called Greenough River Solar Farm, it was built to offset the power requirements of a nearby RO seawater desalination plant and displaces an estimated 20,000 tons of carbon dioxide annually. It does not power the plant directly, however. It was developed by First Solar.

At a small scale, in places where there is a scarcity of fresh, potable groundwater, such as in rural India and Bangladesh, Egypt and the Middle East, off-grid, photovoltaic power may be used in conjunction with different desalination methods. Photovoltaic power is a desirable and logical choice because high solar irradiance is abundant in the regions of India with the greatest desalination needs. However, the economics yet make PV-powered desalination difficult at this scale too. Efforts are ongoing to find different solutions that will work commercially.

One small-scale solution is to deploy an integrated automatic single-axis PV tracking system with programmed tilting angle adjustment, operated during peak sun hours (without batteries) to maximise the production of potable water.

RO is the most prevalent desalination technology used with PVs. It can produce from $1–53 m^3$/day at costs of \$3.6–\$14.9/m^3.[11] A 2.5kW PV system in Rajasthan, India, produces 600 litres of water per hour for six hours each day, reducing total dissolved solids (TDS) from 4,000–6,000 ppm to 450 ppm. The system includes a booster pump, sand filter, RO cartridge and carbon filter.

A PV-RO system exists in Israel to treat groundwater with TDS up to 25,000 ppm and produces 4–5 m^3/day of potable water at <500 ppm TDS per day. It uses variable speed pumps with an efficiency of around 70 per cent. This portable system has the following characteristics:

Feed flow rate: 1.38 m^3/h
Feed pressure: 9.83–33.9 bar
Power consumption: 0.17–1.14 kwh
Permeate TDS: 8.7–160 ppm
Permeate product: 2.46 m^3/h.

Instead of RO, electrodialysis reversal (PV-EDR) desalination may sometimes be more cost-effective. At the groundwater salinity levels found in rural, inland India suffering from brackish groundwater, EDR requires approximately 50 per cent less energy and wastes less water than RO. Research is ongoing in India to optimise a system and design the lowest-cost system for a given location.

6.3.3   *Photovoltaic thermal technology (PV/T)*

This is a new hybrid system which combines the functions of a solar thermal collector and a photovoltaic (PV) panel. PV panels normally heat up on the back when operating, which makes them less efficient. Circulating the brackish water at the back of PV/T panels improves efficiency by cooling the panels. The heated water is then introduced to the desalination module and because it arrives at the RO membrane at a higher temperature this also improves efficiency (unless it is too hot). These systems need to be controlled by automatically varying the fluid flow rate according to the changing circumstances in order to maintain efficiency, guarantee good quality of the produced water and protect the RO membrane by keeping the feed water temperature below the limit tolerated by the membrane.

The latest systems use fuzzy logic controllers to achieve this. These fuzzy logic controllers are already widely used in conventional solar thermal plants to maximise heat production, and to manage the charge delivered in PV systems. Research is ongoing to evaluate the operational temperature changes for each PV/T system segment (glass cover, photovoltaic cells, absorber plate, coolant fluid) and the optimum sizes to achieve the perfect temperature.

6.3.4   *Concentrated evaporation*

This solution is at the pilot stage. It consists of a glazed, insulated metal container which concentrates the sun's heat onto a multiple effect distillation arrangement. Water passed through it evaporates at a much faster rate than would happen otherwise. It uses no other energy inputs. As many of the boxes may be connected together as required to satisfy the local demand. No further information was available from the Israeli manufacturer at the time of writing.

6.3.5 *Plasmon-mediated solar desalination*

This new technology is at the experimental stage. Plasmon-mediated solar desalination utilises an aluminium structure that absorbs photons spanning the 200nm–2,500nm wavelength range, and is claimed by researchers[12] to be both cheap and 'clean'. A plasmon is a quantum of plasma oscillation. Just as light consists of photons, the plasma oscillation consists of plasmons. Plasmons play a large role in the optical properties of metals and semiconductors. Light of frequencies below the plasma frequency is reflected by a material because the electrons in the material screen the electric field of the light. Light of frequencies above the plasma frequency is transmitted by a material because the electrons in the material cannot respond fast enough to screen it. In most metals, the plasma frequency is in the ultraviolet, making them shiny (reflective) in the visible range. The plasmon frequency may occur in the mid-infrared and near-infrared region when semiconductors are in the form of nanoparticles with heavy doping.

Researchers[13] have demonstrated a plasmon-enhanced solar desalination device, fabricated by the self–assembly of aluminium nanoparticles into a three-dimensional porous membrane. This floats naturally on the water surface to be desalinated, efficiently absorbing a broad solar spectrum (>96 per cent) and focusing the absorbed energy at the surface of the water to enable efficient (about 90 per cent) and effective desalination, described as a decrease of four orders of magnitude. These devices are reported to be durable and, with the materials being abundant and of low cost, it has been forecast that, in the future, production could be scalable to provide a portable desalination solution.

6.4   Solar desalination greenhouse

Evaporative cooling and desalination can be combined in hot arid regions by the sea, to create an optimum space for crop growing and clean water provision.

Seawater is drawn in and evaporated by intense solar heat at the front of the greenhouse to create cooler humid conditions inside. Dust, salt spray, pollen and insect pests are trapped and filtered out. The air is cooled by this process, providing good internal climate conditions. Glazing is selective to admit light at wavelengths 430nm and 662nm to optimise photosynthesis and temperature.

Seawater that has not yet evaporated continues through the greenhouse. It is heated by the sun in a network of pipes within the glazing above the growing area, making the air more humid. This air, drawn by fans through the second evaporator, then meets a series of vertical pipes through which cool deep-seawater is passed, whereupon fresh water condenses and trickles down to the base to enter a storage tank. It is then used for irrigation.

There is a commercial installation in Australia. Because of the high capital cost these are suited to growing high-value crops such as tomatoes, which may be harvested all the year

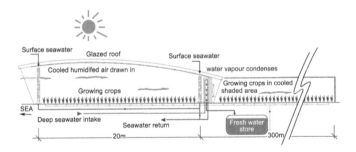

Figure 6.4 Schematic diagram of a solar desalination plant/greenhouse

round despite high temperatures outside because of the internal humidity.

6.5 Small scale: solar stills

At the small, household scale, solar stills may be used for desalination. Thousands of such stills are in operation. Distillation will only remove dissolved salts. If the water also contains suspended solids, a sand filter should be deployed before the water enters the still (See Figure 6.5).

6.5.1 Passive design

A glass or a transparent plastic cover is placed at an angle over a sealed, insulated shallow tray holding the brine with an

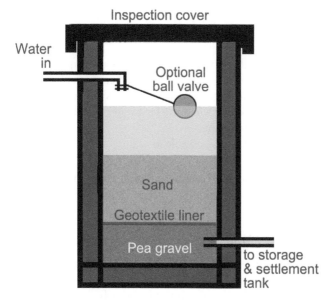

Figure 6.5 A cross-section through a sand filter

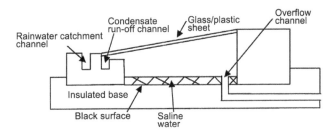

Figure 6.6 Schematic diagram of a single-basin still

absorbent black backing. Where global daily solar irradiation is
$c.5kWh/m^2$ with a still possessing an operational efficiency of
$c.30$ per cent, it may produce $c.2.3L/m^2$ per day. One person's
drinking water requirement is $c.5L/day$. Efficiency is further
maximised with:

1. low-e glazing;
2. the brine depth being 20mm, so it heats more quickly;
3. the collector tilted at $>15°$, ideally perpendicular to the
   midday sun, facing sunwards away from shadows;
4. use in early evening when the feed water is still hot but the
   air temperature falling;
5. keeping the glass and tray clean.

The still can also be used to gather rainwater by adding an external gutter to catch the run-off.

6.5.2 Active PV design

Photovoltaic power should only be used in conjunction with
RO when there is no alternative source of fresh water, as it takes
a great deal of energy to pump water through the membranes.
Figure 6.7 shows a schematic diagram of the required layout,

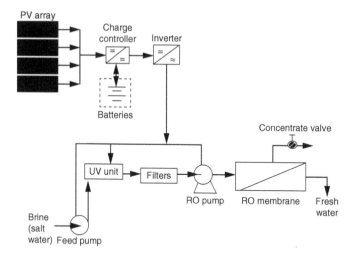

Figure 6.7 Schematic diagram of an AC PV desalination system

which should be sized according to the daily water requirement and the amount of solar insolation.

6.6   Solar drying

Solar drying is used for the drying and preservation of food produce. These three designs ascend in complexity.

6.6.1   Three designs

1. *Basic solar dryer design*: a wooden box with a hinged trans-parent lid, black inside, with an insulated base. Produce is put on a mesh tray or rack above the floor. Air is allowed to flow into the chamber through holes at the front and leave through vents at the top and back.

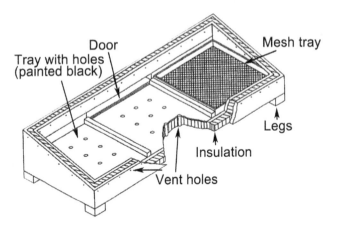

Figure 6.8 Schematic diagram of a box design solar dryer

2. *Box/barn design*: an insulated compartment with a glass cover permits warm air to be drawn in at the bottom, absorbing moisture from produce as it travels via the stack effect to leave near the top. The size (which may be small up to barn-sized) and shape depend upon the products/ amount to be dried. The inside is painted black to absorb the sun's heat. It tilts perpendicular to the midday sun, but at an angle greater than 15° to allow rainwater to run off, and must be situated away from shadows.

3. *Indirect solar dryer design*: air enters through holes at the front of the base. It is drawn by convection. This is created by two glass-covered solar collectors: one just above ground level and one higher up. They both help to draw the air into a vertical cupboard where mesh racks are situated to hold the produce. The air is drawn all around the produce through the mesh. It leaves through a flue at the top. The flue may be painted black to absorb the sun's heat and increase the stack effect. The taller the flue, the more the draw. The produce is not in direct sunlight.

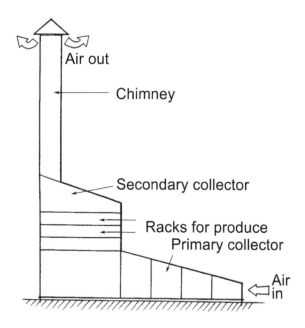

Figure 6.9 Schematic diagram of an indirect solar dryer

6.6.2   *Solar kiln*

A solar kiln is a shed mounted on legs with the sun-facing sides glazed. It is used to dry timber and will dramatically speed up the seasoning process. Air is drawn in through holes in the floor. The air is allowed to leave at the top though an opening that is protected from rain infiltration. The timber is stacked in such a way as to allow the hot air to circulate around it. A solar-powered fan can optionally be used to direct the air down to the timber (See Figure 6. 10).

6.7   **Solar pasteurisation**

Pasteurisation of a liquid (water or milk) – that may contain harmful germs, viruses and parasites including cholera and

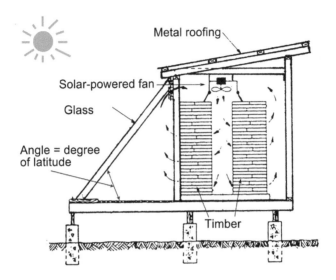

Figure 6.10 Solar kiln cross-section

hepatitis A and B – will kill them all. Pasteurisation requires heating to 65°C (149°F) for six minutes to remove these pathogens. Boiling is unnecessary. Milk is pasteurised at 71°C (160°F) for 15 seconds. Solar energy may be used for this process. This removes the need to use expensive or polluting fuels.

A solar box heater is used to heat the water. This waterproof box is black inside and covered with glass. It is then connected to the polluted water supply and placed in the sun. A water pasteurisation indicator (WAPI) is used to indicate when the required temperature has been reached. A WAPI is a clear plastic tube partially filled with a soybean wax that melts at about 70°C (158°F). With the solid wax at the top end of the tube, the WAPI is placed in the bottom of the solar box heater. If the wax melts and falls to the bottom of the tube, it indicates that water pasteurisation conditions have been reached.

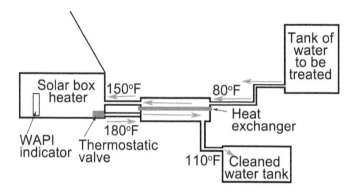

Figure 6.11 Schematic diagram of the solar water pasteurisation device

In the schematic diagram for a design for a small-scale treatment system (Figure 6.11), polluted liquid is directed across a thin metal plate heat exchanger into the solar heater. This water absorbs some heat from the treated water that is leaving the glazed box beneath the metal plate. This preheats it before it enters the glazed box, making it more efficient. A thermostatic valve only allows the water to leave the cooker when it has reached the required temperature and been treated successfully. The WAPI indicator should be placed at the bottom of the box to indicate pasteurisation conditions have been achieved.

Notes

1 World Bank, *Renewable Energy Desalination: An Emerging Solution to Close the Water Gap in the Middle East and North Africa*, World Bank, Washington, DC (2012). doi: 10.1596/978-0-8213-8838-9. Licence: Creative Commons Attribution CC BY 3.0.

2 Ibid.

3 Ibid.

4 Fichtner and DLR. 2011. MENA Regional Water Outlook, Part II, Desalination Using Renewable Energy, Task 1–Desalination

Potential; Task 2–Energy Requirements; Task 3–Concentrate Management. Final Report, commissioned by the World Bank, Fichtner and DLR. www.worldbank.org/mna/watergap.

5 Ibid.

6 Ibid.

7 Trieb, F., *et al.*, 'Global potential of concentrating solar power', presented at SolarPaces Conference Berlin, September (2009), available at: http://bit.ly/2sLWv2u

8 World Bank (2012).

9 DLR, *Concentrating Solar Power for the Mediterranean Region*, DLR, Stuttgart (2005).

10 Ibid.

11 Al-Karaghouli, A., Renne, D., and Kazmerski, L.L., 'Solar and wind opportunities for water desalination in the Arab regions', *Renewable and Sustainable Energy Reviews*, 13, 2397–407 (2009).

12 Tianyu Liu and Yat Li, 'Photocatalysis: plasmonic solar desalination', *Nature Photonics*, 10, 361–62 (2016). http://go.nature.com/2eMPDjt

13 Lin Zhou *et al.*, '3D self-assembly of aluminium nanoparticles for plasmon-enhanced solar desalination', *Nature Photonics*, 10, 393–8 (2016). http://go.nature.com/2fnou2Q

7   Solar cooling

Commercial, large residential and industrial buildings may use active solar technologies for ventilation air preheating, solar process heating and solar cooling. Solar power is a good match for cooling/air-conditioning, because the heat is available when most needed, but commercial deployment is at an early stage.

7.1   Space cooling

Space cooling uses thermally activated cooling systems driven (or partially driven) by solar energy. The two systems are:

1. *closed cycle*: a heat-driven heat pump that operates in a closed cycle with a working fluid pair, usually an *absorbent*-refrigerant such as LiBr-water and water-ammonia, or an *adsorption* cycle using sorption such as silica gel; two or more absorbers are used to continuously provide chilled water;
2. *open cycle*: solar thermal energy regenerates *desiccant substances* such as water by drying them, thereby cooling the air. Liquid or solid desiccants are possible, providing a combination of dehumidification and evaporative cooling of air.

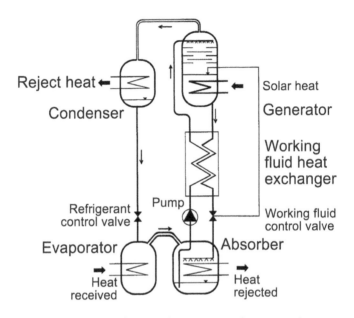

Figure 7.1 Schematic diagram of a continuous absorption refrigeration process

7.1.1 *Absorption NH₃/H₂O*

The single-stage, continuous absorption refrigeration process works as follows: the working fluid (WF), mainly ammonia and water, is boiled in the generator. This receives heat from the solar collectors at 65–80°C. Mainly ammonia, but some water, leaves at the top and is condensed in the water-cooled condenser (25–35°C).

The boiling working fluid in the generator has therefore to be exchanged continuously to deliver a strong working fluid with a concentration of 40 per cent ammonia. This is pumped from the absorber via the working fluid heat exchanger. The heat exchanger heats it to 50–65°C, using heat exchanged from the weaker fluid leaving the generator. The latter, now cooler, is led to the absorber, from which it leaves at c.35°C.

Meanwhile, the condensed refrigerant ammonia has left the condenser and is injected into the evaporator by the refrigerant control valve. This works at low pressure level (2–4 bar), to make the refrigerant boil and evaporate. The cold vapour flows into the absorber which absorbs it, combines it with the working fluid and sends in back to the generator.

The thermal coefficient of performance of this type of unit ($COP_{thermal}$) describes the relation between the benefit (cooling capacity) and the input (heat from the collectors):

$$COP_{thermal} = Q_{cooling}/Q_{heating}$$

7.1.2   *Absorption H_2O/LiBr*

This process employs a refrigerant expanding from a condenser to an evaporator through a throttle, in an absorber/desorber combination that is akin to a 'thermal compressor' in a conventional vapour compression cycle. Cooling is produced through the evaporation of the refrigerant (water) at low temperature. The absorbent then absorbs the refrigerant vapour at low pressure and desorbs into the condenser at high pressure when (solar) heat is supplied (see Figure 7.2).

To describe this in more detail: a single-effect absorption system liquid refrigerant leaves the condenser through the throttle valve into the evaporator. It expands as it does so, taking its heat of evaporation from the stream of chilled water in the absorber and cooling the latter. The vapour leaving is absorbed by an absorbent solution. This arrives dilute in refrigerant (strong absorption capability) and it leaves rich in refrigerant (weak absorption capability). It is pumped via a heat exchanger to a desorber which, by applying heat from the solar-heated water stream, causes the release of water and regenerates the

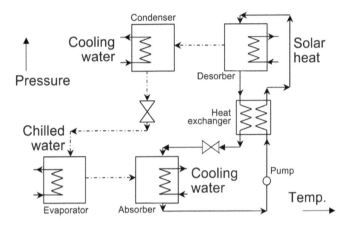

Figure 7.2 Schematic diagram of a H_2O/LiBr absorption refrigeration process

refrigerant solution to a strong state. This proceeds to the condenser where it condenses to liquid, then the cycle begins again as it goes on to expand into the evaporator. Both the absorber and condenser are cooled by streams of cooling water to reject the heats of absorption and condensation respectively.

7.1.3 Adsorption

Adsorption substances are working pairs, usually water/silica gel. The solid sorbent (gel) is alternately cooled and heated to be able to adsorb and desorb the refrigerant (water). A sequence of adsorbers is deployed to use the heat from one to power another. The cycle is as follows (refer to Figure 7.3): refrigerant previously adsorbed in one adsorber is driven off through the use of hot water (may be solar-heated) (right compartment). It then condenses in the condenser and the heat of condensation is removed by cooling water. The condensate is sprayed in the evaporator and evaporates under low partial pressure, producing

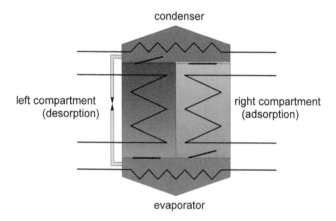

condenser

left compartment
(desorption)

right compartment
(adsorption)

evaporator

Figure 7.3 Schematic diagram of an adsorption refrigeration process

cooling power. The refrigerant vapour is adsorbed into the other adsorber (left) where heat is removed by cooling water.

7.1.4 Open cycle liquid desiccant cooling

This process is used to cool and dehumidify air (refer to Figure 7.4). Water vapour is removed from the air by a desiccant, which is then regenerated (dried) by heat from an available source – which can be solar. Both solid and liquid hygroscopic materials may be used as the desiccant. Liquid desiccant systems can store cooling capacity when the water vapour is condensed and stored. Solar thermal energy is used whenever available to run the desorber and its associated components (hot water-to-solution heat exchanger, air-to-air recuperator, pump) to concentrate hygroscopic salt. Later, when needed, this is used to dehumidify process air. This method of cold storage is the most compact, requires no insulation and can be applied for indefinitely long time periods.

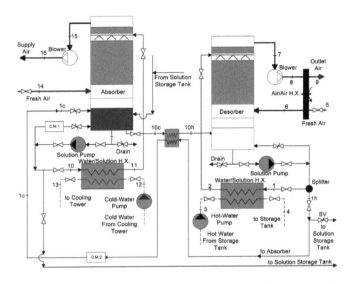

Figure 7.4 Schematic diagram of an open cycle liquid desiccant cooling system

7.1.5 Solid desiccant air handling unit

Two air channels are mounted on top of each other. The outdoor air enters (A) where the sorption wheel with a silica gel surface dehumidifies is (B) and transfers heat from the outgoing air (C), rehumidifies it to the correct level then enters the conditioned space, increases its enthalpy by internal heat sources and moisture, and leaves as return air (G), where moisture (H and J) and heat (I) are removed as necessary and it is expelled (L). Highly effective solar collectors should be used for the heat regeneration. The middle European climate allows an enhancement of the adiabatic cooling mode.

Relation between the cooling capacity and the regeneration heat:

$$COP_{thermal,\ plant} = Q_{c,plant}/Q_{heat}$$

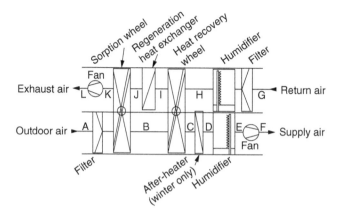

Figure 7.5 Schematic diagram of a solid desiccant air handling system

Relation between the cooling load and the regeneration heat:

$$COP_{thermal, build} = Q_{c, build}/Q_{heat}$$

7.1.6 *Desiccant-enhanced evaporative (DEVAP) air-conditioner*

NREL, AILR Research, Inc. and Synapse Product Development have developed the DEVAP air-conditioner. This consists of two stages: dehumidifier and indirect evaporative cooling. Water is added to the tops of both; liquid desiccant is pumped through the first. Some outdoor air is mixed with return air from the building to form the supply air stream, which flows left to right through the two stages. In the dehumidifier, a membrane contains the desiccant while humidity from the supply air passes through it to the desiccant, which is also in thermal contact with a flocked, wetted surface that is cooled as outdoor air passes by it, causing the water to evaporate and indirectly cooling the desiccant. In

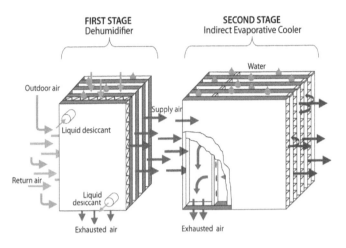

Figure 7.6 Schematic diagram of a DEVAP system

stage 2, the supply air passes by a water-impermeable surface that is wetted and flocked on its opposite side, providing indirect evaporative cooling. A small fraction of the cool, dry supply air is then redirected through the second-stage evaporative passages to evaporate water from the flocked surface and is then exhausted.

7.2 Solar design considerations

The performance of a thermally driven chiller is described by the thermal coefficient of performance (COP). This is defined as the produced cold per unit of driving heat. Single-effect absorption systems are limited in COP to about 0.7. Therefore they require a rather large solar collector area to supply the heat needed for their operation. Double effect can reach a COP up to 1.2 and triple-effect up to 1.7. A detailed description of absorption systems is available in the ASHRAE *Handbook*.

The goal of the design is to minimise the use of fossil fuels required for backup power of the chiller. This means minimising the load with passive solar techniques, using a large enough solar collector area for the load and integrating a large enough heat dump.

It is worth remembering that electrical power is also required for the pumps and/or fans of the solar collector circuits and those of the chillers. These may be driven by a PV system. Desiccant chiller systems require the highest electricity consumption because of their high overall pressure drop.

The economic use of the solar collector should be maximised by also letting it supply heat to other loads such as space heating or hot water when required (e.g. in winter).

For closed cycle (absorption and adsorption) chillers, sophisticated controls are required to match the loads with the available solar heat from the collectors, since each possesses a reverse relationship of efficiency to the ambient temperature. There is a 'sweet point' where the best COP is obtained, which depends on

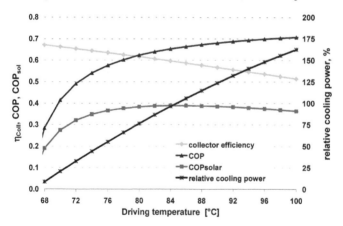

Figure 7.7 The relationship between collector efficiency η_{Coll}, COP, COP$_{solar}$ and cooling power (relative) and the operation temperature of a solar collector and a thermally driven chiller. This example is for a single-effect absorption chiller and flat plat collector with incident radiation on the collector at 800 W/m^2, a cooling water temperature of 29°C, and a chilled water temperature of 9°C.

Source: SOLAIR

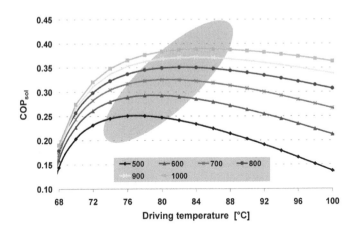

Figure 7.8 The relationship between COP$_{solar}$ and the operation temperature of a solar collector and a thermally driven chiller. This example is for a single-effect absorption chiller and flat plat collector with incident radiation on the collector of 500–1000 W/m², a cooling water temperature of 29°C and a chilled water temperature of 9°C. The marked area indicates the operation conditions leading to highest COP$_{solar}$ values

Source: SOLAIR

the changing demands of the load and the amount of irradiation. This is illustrated in Figures 7.7 and 7.8. Advanced controls would detect and match the cooling power of the system to the load, adjusting the supply temperature within the range available to the optimum that provides the most efficient cooling service.

7.2.1 *Evaluation*

More information and a simplified evaluation tool called 'Easy Solar Cooling' can help assess the cost performance of different technologies and system designs under different operating conditions. For this and plenty of case studies, see: http://www.solair-project.eu/218.0.html

8 Making the business case

A solar installation may well save money over its lifetime but have a relatively high initial capital cost. To make the business case for the investment it is necessary to show its lifecycle value – in financial and other terms. The financial value of the project is usually assessed by calculating the total costs and benefits either over its lifetime or over a set investment period, using discounted cash flow (DCF), and a comparison of the result with other forms of investment.

Other factors to be brought into play might be: carbon emission and other pollution savings over the lifetime of the project; ethical considerations; co-benefits such as improved air quality and health; the availability of tax credits, grant-funding and feed-in tariffs; and energy security or the uncertainty of future fuel costs.

8.1 Financial case

There are four stages in calculating DCF.

8.1.1 Estimate the resulting cash flow

The cash flow is taken from the estimated savings in energy cost resulting from the measure taken. This will depend upon projections of future energy cost. For example, energy prices over the last three years can be projected on a median basis into the future. You might also factor in external savings on, for example, workers' days off due to ill health saved by a measure that will improve their health, or credits for reducing greenhouse gas emissions.

This will then need to be discounted at a discount rate to be chosen. The discount rate takes account of the future value of money in today's prices, given an assumed rate of inflation. An average price (P) is calculated this way for each year of the projected lifetime (L) of the project. Each of these figures is then multiplied by the amount of energy (E) expected to be saved every year.

The lifetime period chosen for the project will depend upon the expected lifetime of the technology. If it were a PV panel, for example, it could be 25 years. Should it be an insulation measure, it could be 35 years or more. The total cost savings (S) generated by energy not used compared to not doing the project, over the lifetime of the project will then be:

$$S = E \times [P(\text{year 1})] + E \times [P(\text{year 2})] + E \times [P(\text{year 3})] \ldots$$
$$E \times [P(\text{year L})]$$

8.1.2 Apply the discount rate

The chosen discount rate is subjective. The industrial model ENUSIM uses private fuel prices and a 10 per cent discount rate to reflect the incentives faced by firms. Some organisations adopt the rate used in their government's modelling. Others adopt the current rate of inflation, or interest rate on a loan taken out for the purpose of the measure that would need to be repaid.

If a discount rate of 3.5 per cent per annum is chosen, this implies that it values £1 today equally with £1.035 in a year's time; or that £1 in a year's time is worth only 96.62p now, because 1/1.035 equals 0.9662. The 96.62p figure is called the present value (PV) of the £1, and the 0.9662 figure is the 'discount factor'.

Table 8.1 Discount factors at 1–10 per cent over 30 years

Year	Discount factors (%)										
	1.0	2.0	3.0	3.5	4.0	5.0	6.0	7.0	8.0	9.0	10.0
0	1.0000	1.0000	1.0000	1.0000	1.0000	1.0000	1.0000	1.0000	1.0000	1.0000	1.0000
1	0.9901	0.9804	0.9709	0.9662	0.9615	0.9524	0.9434	0.9346	0.9259	0.9174	0.9091
2	0.9803	0.9612	0.9426	0.9335	0.9246	0.9070	0.8900	0.8734	0.8573	0.8417	0.8264
3	0.9706	0.9423	0.9151	0.9019	0.8890	0.8638	0.8396	0.8163	0.7938	0.7722	0.7513
4	0.9610	0.9238	0.8885	0.8714	0.8548	0.8227	0.7921	0.7629	0.7350	0.7084	0.6830
5	0.9515	0.9057	0.8626	0.8420	0.8219	0.7835	0.7473	0.7130	0.6806	0.6499	0.6209
6	0.9420	0.8880	0.8375	0.8135	0.7903	0.7462	0.7050	0.6663	0.6302	0.5963	0.5645
7	0.9327	0.8706	0.8131	0.7860	0.7599	0.7107	0.6651	0.6227	0.5835	0.5470	0.5132
8	0.9235	0.8535	0.7894	0.7594	0.7307	0.6768	0.6274	0.5820	0.5403	0.5019	0.4665
9	0.9143	0.8368	0.7664	0.7337	0.7026	0.6446	0.5919	0.5439	0.5002	0.4604	0.4241
10	0.9053	0.8203	0.7441	0.7089	0.6756	0.6139	0.5584	0.5083	0.4632	0.4224	0.3855
11	0.8963	0.8043	0.7224	0.6849	0.6496	0.5847	0.5268	0.4751	0.4289	0.3875	0.3505
12	0.8874	0.7885	0.7014	0.6618	0.6246	0.5568	0.4970	0.4440	0.3971	0.3555	0.3186
13	0.8787	0.7730	0.6810	0.6394	0.6006	0.5303	0.4688	0.4150	0.3677	0.3262	0.2897
14	0.8700	0.7579	0.6611	0.6178	0.5775	0.5051	0.4423	0.3878	0.3405	0.2992	0.2633
15	0.8613	0.7430	0.6419	0.5969	0.5553	0.4810	0.4173	0.3624	0.3152	0.2745	0.2394

(Continued)

Table 8.1 Discount factors at 1–10 per cent over 30 years (Continued)

Year	Discount factors (%)										
	1.0	2.0	3.0	3.5	4.0	5.0	6.0	7.0	8.0	9.0	10.0
16	0.8528	0.7284	0.6232	0.5767	0.5339	0.4581	0.3936	0.3387	0.2919	0.2519	0.2176
17	0.8444	0.7142	0.6050	0.5572	0.5134	0.4363	0.3714	0.3166	0.2703	0.2311	0.1978
18	0.8360	0.7002	0.5874	0.5384	0.4936	0.4155	0.3503	0.2959	0.2502	0.2120	0.1799
19	0.8277	0.6864	0.5703	0.5202	0.4746	0.3957	0.3305	0.2765	0.2317	0.1945	0.1635
20	0.8195	0.6730	0.5537	0.5026	0.4564	0.3769	0.3118	0.2584	0.2145	0.1784	0.1486
21	0.8114	0.6598	0.5375	0.4856	0.4388	0.3589	0.2942	0.2415	0.1987	0.1637	0.1351
22	0.8034	0.6468	0.5219	0.4692	0.4220	0.3418	0.2775	0.2257	0.1839	0.1502	0.1228
23	0.7954	0.6342	0.5067	0.4533	0.4057	0.3256	0.2618	0.2109	0.1703	0.1378	0.1117
24	0.7876	0.6217	0.4919	0.4380	0.3901	0.3101	0.2470	0.1971	0.1577	0.1264	0.1015
25	0.7798	0.6095	0.4776	0.4231	0.3751	0.2953	0.2330	0.1842	0.1460	0.1160	0.0923
26	0.7720	0.5976	0.4637	0.4088	0.3607	0.2812	0.2198	0.1722	0.1352	0.1064	0.0839
27	0.7644	0.5859	0.4502	0.3950	0.3468	0.2678	0.2074	0.1609	0.1252	0.0976	0.0763
28	0.7568	0.5744	0.4371	0.3817	0.3335	0.2551	0.1956	0.1504	0.1159	0.0895	0.0693
29	0.7493	0.5631	0.4243	0.3687	0.3207	0.2429	0.1846	0.1406	0.1073	0.0822	0.0630
30	0.7419	0.5521	0.4120	0.3563	0.3083	0.2314	0.1741	0.1314	0.0994	0.0754	0.0573

Discounting is conducted at annual intervals. It is useful to run the calculation several times with different discount rates. Here is a table of discount factors at 1–10 per cent over 30 years:

8.1.3 Calculate the net present value (NPV)

From the figure for the total cost savings (S) is deducted the discounted present cost (C) of the measure. This gives the net present value (NPV) of the project, representing the value in today's money of the net profit (or loss) that will be generated in the future.

$$NPV = S - C$$

This is the most useful way of comparing the value of different measures because it accounts for the full value of the project and presents it in easily comparable form.

8.1.4 Compare the internal rate of return (IRR)

The NPV lets the projects be compared to what would happen to the same amount of money were it to be invested in a bank account with the same interest rate as the discount rate chosen, by calculating the internal rate of return (IRR). Refer to Figure 8.1. Using Microsoft Excel:

1. type the initial expenditure into a cell as a negative number;
2. enter the subsequent discounted cash return figures for each year into the cells beneath this;
3. reveal the IRR by typing into the cell beneath this list in the column the function command: '=IRR(A1:A4)', where the cell range 'A1:A4' is replaced by the cell range you have used. Then press the enter key. In Figure 8.1, the IRR value, in this case 18 per cent, is then displayed in that cell.

A5	▲▼	⊗	◀	⦁ *fx*	=IRR(A1:A4)

	A	B	C	D	E
1	−60000				
2	30000				
3	27000				
4	24300				
5	18%				

Figure 8.1 Using Microsoft Excel to calculate the internal rate of return of an investment. The formula in the field at the top is entered into cell A5 and yields the IRR based on the figures above

This exercise should then be repeated for alternative investment strategies and the results compared.

Further ways of offsetting risk are: asset finance such as leasing and renting – for example, leasing land or roof space to a local utility or energy firm to generate electricity from photovoltaic panels.

8.2   Carbon offsetting

Carbon offsetting is the counteracting of carbon dioxide or other greenhouse gas emissions from one activity with an equivalent reduction of emissions to the atmosphere elsewhere. It is a controversial but sometimes useful practice. Carbon offsetting projects may include the prevention of emissions by energy efficiency activities, the installation of renewable energy generators, the planting of trees or the use of any means to absorb and safely and permanently sequester carbon dioxide from the atmosphere. The project must not have occurred in any case but be additional to that which would have happened without the offsetting action. It must be measurable and follow a comprehensive set of validation and verification procedures to demonstrate effectiveness on a regular basis using independent third parties.

A given solar power installation may achieve credits to assist with project financing under a UN/World Bank-brokered financing scheme, or under a third-party scheme where a business purchases 'carbon offsets' from an intermediary. Or a solar project may simply want to calculate the credits to otherwise justify itself. The energy generated by the project over a period of time can be estimated and claimed to offset the equivalent amount of energy from a fossil fuel source that it might displace or prevent the use of. It will then have saved the emissions equivalent to those arising from burning that amount of fuel – depending upon the fuel chosen. Different fuels emit differing amounts of greenhouse gases when burnt, per unit of energy generated. These may be calculated in $kgCO_2e$ (the 'e' means that all greenhouse gases are converted to carbon dioxide equivalence in warming terms) by using the values in Table 8.2.

Other greenhouse gases emitted by the burning of fossil fuels, such as CO_2, CH_4 and N_2O, are converted for convenience into their equivalent in terms of damage caused by global warming – 'carbon dioxide equivalent' units or CO_2e. The greenhouse gas conversion factor is quoted as $kgCO_2e$ per unit of fuel consumed.

If the fuel is electricity the conversion factor varies according to the generation technology mix (the amount of coal, gas, nuclear, renewables and oil) in the local grid.

Table 8.2 Carbon dioxide emission factors by gross calorific value

Energy source	$KgCO_2/kWh$	$KgCO_2$ per other units
Natural gas	0.18523	5.3808 per therm
LPG	0.21445	6.2915 per therm
Coal	0.32227	2,383 per tonne
Diesel	0.25301	3,188 per tonne
Petrol	0.24176	2.3117 per litre
Fuel oil	0.26592	3,228 per tonne
Burning oil	0.24683	3,165 per tonne
Wood pellets	0.03895	183.9 per tonne

The benefit has been estimated at €12/MWh in Europe, compared with €23/MWh at the global scale, based on a global average GHG emission factor from electricity production of 0.6 $kgCO_2$/kWh. This includes 12–25gCO_2/kWh emitted from the PV lifecycle and assumes a CO_2 abatement cost of €20/tCO_2. This estimate is conservative because the cost of CO_2 abatement in fossil fuel power plants is likely to be well above €20/tCO_2 in the long term. Other cost benefits that are external to a solar project include: the reduction of grid losses due to distributed generation (on the order of €5/MWh); the positive impact on energy security (€15–30/MWh, depending on fossil fuel prices) and on electricity demand peaks (i.e. peak shaving), thus reducing the need for additional peak capacity (€10/MWh).

9 Units

9.1 SI radiometric units

Table 9.1 SI radiometric units

Quantity		Unit		Dimension M = mass L = length T = time	Notes
Name	Symbol	Name	Symbol	Symbol	
Energy	Q	joule	J	$\mathbf{M} \cdot \mathbf{L} \cdot \mathbf{T}$	energy
Power	Φ	watt	W *or* J/s	$\mathbf{M} \cdot \mathbf{L}^2 \cdot \mathbf{T}^{-3}$	radiant energy per unit time, also called *radiant power*
Spectral power	Φ_λ	watt per metre	$\mathrm{W} \cdot \mathrm{m}^{-1}$	$\mathbf{M} \cdot \mathbf{L} \cdot \mathbf{T}^{-3}$	radiant power per wavelength
Radiant intensity	I_e	watt per steradian	$\mathrm{W} \cdot \mathrm{sr}^{-1}$	$\mathbf{M} \cdot \mathbf{L}^2 \cdot \mathbf{T}^{-3}$	power per unit solid angle
Spectral intensity	$I_{e\lambda}$	watt per steradian per metre	$\mathrm{W} \cdot \mathrm{sr}^{-1} \cdot \mathrm{m}^{-1}$	$\mathbf{M} \cdot \mathbf{L} \cdot \mathbf{T}^{-3}$	radiant intensity per wavelength

(Continued)

Table 9.1 SI radiometric units (*Continued*)

Radiance (also called 'radiant flux intensity', RFD)	L_e	watt per steradian per square metre	$W \cdot sr^{-1} \cdot m^{-2}$	$M \cdot T^{-3}$	power per unit solid angle per unit *projected* source area
Spectral radiance	$L_{e\lambda}$ or $L_{e\nu}$	watt per steradian per metre³ *or* watts per steradian per square metre per hertz	$W \cdot sr^{-1} \cdot m^{-3}$ *or* $W \cdot sr^{-1} \cdot m^{-2} \cdot Hz^{-1}$	$M \cdot L^{-1} \cdot T^{-3}$ *or* $M \cdot T^{-2}$	commonly measured in $W \cdot sr^{-1} \cdot m^{-2} \cdot nm^{-1}$ with surface area and either wavelength or frequency
Irradiance	E_e	watt per square metre	$W \cdot m^{-2}$	$M \cdot T^{-3}$	power incident on a surface, also called RFD
Spectral irradiance	$E_{e\lambda}$ or $E_{e\nu}$	watt per metre³ *or* watts per square metre per hertz	$W \cdot m^{-3}$ *or* $W \cdot m^{-2} \cdot Hz^{-1}$	$M \cdot L^{-1} \cdot T^{-3}$ *or* $M \cdot T^{-2}$	commonly measured in $W \cdot m^{-2} \cdot nm^{-1}$ or 10^{-22} $W \cdot m^{-2} \cdot Hz^{-1}$, known as solar flux unit
Radiant exitance/ radiant emittance	M_e	watt per square metre	$W \cdot m^{-2}$	$M \cdot T^{-3}$	power emitted from a surface
Spectral radiant exitance/ spectral radiant emittance	$M_{e\lambda}$ or $M_{e\nu}$	watt per metre³ *or* watt per square metre per hertz	$W \cdot m^{-3}$ *or* $W \cdot m^{-2} \cdot Hz^{-1}$	$M \cdot L^{-1} \cdot T^{-3}$ *or* $M \cdot T^{-2}$	power emitted from a surface per unit wavelength or frequency

(*Continued*)

Table 9.1 SI radiometric units (*Continued*)

Radiosity	J_e	watts per square metre	$W \cdot m^{-2}$	$\mathbf{M \cdot T^{-3}}$	emitted plus reflected power leaving a surface
Spectral radiosity	$J_{e\lambda}$	watts per metre^3	$W \cdot m^{-3}$	$\mathbf{M \cdot L^{-1} \cdot T^{-3}}$	emitted plus reflected power leaving a surface per unit wavelength
Radiant exposure	H_e	joules or watt-hours per square metre	$J \cdot m^{-2}$ or Wh/m^2	$\mathbf{M \cdot T^{-2}}$	also referred to as fluence or irradiation
Radiant energy density	ω_e	joules per metre^3	$J \cdot m^{-3}$	$\mathbf{M \cdot L^{-1} \cdot T^{-2}}$	

9.2   Prefixes

milli-	m	10^{-3}	
kilo-	k	10^3	1,000
mega-	M	10^6	1,000,000
giga-	G	10^9	1,000,000,000
tera-	T	10^{12}	1,000,000,000,000
peta-	P	10^{15}	1,000,000,000,000,000

For example:

milliwatt (mW): 1000th of a watt
kilowatt (kW): 1,000W
megawatt (MW): 1,000,000W

gigawatt (GW): 1,000,000,000W

terawatt (TW): 1,000,000,000,000W. In 2006 about 16TW of
  power was used worldwide.

MJ = megajoule

TJ = terajoule

GWh = gigawatt-hours

9.3   Other energy units

Btu = British thermal unit (MBtu = millions of Btus)

toe = tonnes of equivalent oil (Mtoe = millions of toe)

cal = the gram calorie: the approximate amount of energy
  needed to raise the temperature of one gram of water by
  one degree Celsius at a pressure of one atmosphere. Not to
  be confused with Cal = the kilogram calorie: the food
  calorie, equal to 1,000 small calories, 1 kilocalorie
  (symbol: kcal).

9.4   Conversion factors

Table 9.2 Conversion factors for some units of energy

To:	TJ	Gcal	Mtoe	MBtu	GWh
From:	Multiply by:				
terajoule (TJ)	1	238.8	2.388×10^{-5}	947.8	0.2778
gigacalorie (Gcal)	4.1868×10^{-3}	1	10^{-7}	3.968	1.163×10^{-3}
million tonne of oil equivalent (Mtoe)	4.1868×10^{4}	10^{7}	1	3.968×10^{7}	11,630
million British thermal unit (MBtu)	1.0551×10^{-3}	0.252	2.52×10^{-8}	1	2.931×10^{-4}
gigawatt hour (GWh)	3.6	860	8.6×10^{-5}	3,412	1

(*Continued*)

Table 9.2 Conversion factors for some units of energy (*Continued*)

From	to kWh. Multiply by:
Therms	29.31
Btu	2.931×10^{-4}
MJ	0.2778
Toe	1.163×10^{4}
Kcal	1.163×10^{-3}

Example:

- Conversion of 100,000 Btu to kWh:
- 100,000 Btu = 100,000 × 2.931 × 10^{-4} kWh = 29.31kWh

9.4.1 Conversion factors for mass

Table 9.3 Conversion factors for some units of mass

To:	kg	T	lt	st	lb
From:	Multiply by:				
kilogram (kg)	1	0.001	9.84×10^{-4}	1.102×10^{-3}	2.2046
tonne (t)	1,000	1	0.984	1.1023	2204.6
long ton (lt)	1,016	1.016	1	1.120	2240.0
short ton (st)	907.2	0.9072	0.893	1	2000.0
pound (lb)	0.454	4.54×10^{-4}	4.46×10^{-4}	5.0×10^{-4}	1

9.4.2 Conversion factors for volume

Table 9.4 Conversion factors for some units of volume

To:	gal US	gal UK	bbl	ft^3	l	m^3
From:	Multiply by:					
US gallon (gal)	1	0.8327	0.02381	0.1337	3.785	0.0038
UK gallon (gal)	1.201	1	0.02859	0.1605	4.546	0.0045
barrel (bbl)	42.0	34.97	1	5.615	159.0	0.159
cubic foot (ft^3)	7.48	6.229	0.1781	1	28.3	0.0283
litre (l)	0.2642	0.220	0.0063	0.0353	1	0.001
cubic metre (m^3)	264.2	220.0	6.289	35.3147	1000.0	1

9.5   Power and energy

Power is the rate at which energy is produced by a generator or
    consumed by an appliance.

Unit: the watt (W). 1,000 watts is a kilowatt (kW).

Energy is the amount of power produced by a generator or
    consumed by an appliance or over a period of time.

Unit: the watt-hour (Wh). 1,000 watt-hours is a kilowatt-hour
    (kWh), commonly a unit of electricity on a bill.

Alternate unit: the joule (J). Watt-hours can be used to describe
    heat energy as well as electrical energy, but joules are also
    used for heat. 3,600 Joules = 1Wh. A joule is one watt
    per second, since there are 3,600 seconds in an hour; or
    3.6 megajoules (MJ) = 1kWh.

Example:

One photovoltaic solar panel producing 80W for two hours,
or two panels producing 80W for one hour would produce
$2 \times 80 = 160$Wh.

   Three panels producing 90W for five hours will produce
$3 \times 90 \times 5 = 1350$Wh or 1.35kWh.

10 Standards

The following are some of the relevant international standards for the field of solar energy. They are set by two main bodies.

10.1 Standard setting bodies

The International Organization for Standardization (ISO) is an international standard setting body composed of representatives from various national standards organisations. For more information see: www.iso.org

The International Electrotechnical Commission (IEC) is a global organisation for the preparation and publication of International Standards for electrical, electronic and related technologies. Its Technical Committee (TC) 82 prepares International Standards for systems of photovoltaic conversion of solar energy into electrical energy and for all the elements in the entire photovoltaic energy system. It has prepared standards for terms and symbols, salt mist corrosion testing, design qualification and type approval of crystalline silicon and thin film modules, and characteristic parameters of stand-alone systems, among others. In the future, TC 82 work will include:

- system commissioning, maintenance and disposal;
- characterisation and measurement of new thin film photovoltaic module technologies such as CdTe, CIS, $CuInSe_2$ and so forth;

- new technology storage systems;
- applications with special site conditions, such as tropical zone, northern latitudes and marine areas.

For more information see: www.iec.ch

10.2   Standards for solar measurement

- ISO 21348:2007: Space environment (natural and artificial) – Process for determining solar irradiances
- ISO 9022-9:1994: Optics and optical instruments – Environmental test methods – Part 9: Solar radiation
- ISO 9059:1990: Solar energy – Calibration of field pyrheliometers by comparison to a reference pyrheliometer
- ISO 9060:1990: Solar energy – Specification and classification of instruments for measuring hemispherical solar and direct solar radiation
- ISO 9488:1999: Solar energy – Vocabulary
- ISO 9845-1:1992: Solar energy – Reference solar spectral irradiance at the ground at different receiving conditions – Part 1: Direct normal and hemispherical solar irradiance for air mass 1,5
- ISO 9846:1993: Solar energy – Calibration of a pyranometer using a pyrheliometer
- ISO 9847:1992: Solar energy – Calibration of field pyranometers by comparison to a reference pyranometer
- ISO/TR 17801:2014: Plastics – Standard table for reference global solar spectral irradiance at sea level – Horizontal, relative air mass 1
- ISO/TR 9901:1990: Solar energy – Field pyranometers – Recommended practice for use.

10.3   Standards for solar thermal

- ISO 22975-3:2014: Solar energy – Collector components and materials – Part 3: Absorber surface durability

- ISO 9459-1:1993: Solar heating – Domestic water heating systems – Part 1: Performance rating procedure using indoor test methods
- ISO 9459-2:1995: Solar heating – Domestic water heating systems – Part 2: Outdoor test methods for system performance characterization and yearly performance prediction of solar-only systems
- ISO 9459-4:2013: Solar heating – Domestic water heating systems – Part 4: System performance characterization by means of component tests and computer simulation
- ISO 9459-5:2007: Solar heating – Domestic water heating systems – Part 5: System performance characterization by means of whole-system tests and computer simulation
- ISO 9553:1997: Solar energy – Methods of testing preformed rubber seals and sealing compounds used in collectors
- ISO 9806:2013: Solar energy – Solar thermal collectors – Test methods
- ISO 9808:1990: Solar water heaters – Elastomeric materials for absorbers, connecting pipes and fittings – Method of assessment
- ISO/TR 10217:1989: Solar energy – Water heating systems – Guide to material selection with regard to internal corrosion.

10.4 Standards for solar PV

- IEC 61215 covers crystalline types
- IEC 61646 covers thin film types
- IEC 61730: Photovoltaic module safety qualification
- Modules must also carry a CE mark
- ISO 15387:2005: Space systems – Single-junction solar cells – Measurements and calibration procedures.

Note: solar PV installation standards are set on a national basis. Consult the relevant authority.

11 Resources for calculation and modelling

11.1 Datasets

- NOAA Sunrise/Sunset and Solar Position Calculators and spreadsheets: http://1.usa.gov/1g8xFme
- NREL data: http://www.nrel.gov/gis/data_solar.html and www.nrel.gov/rredc/pvwatts
- Calendar of sunrise, sunset, noon daylight at any location for an entire year. The table shows the time and azimuth in degrees: http://bit.ly/1njenNl
- Sun position calculator producing sun path diagrams from location found using Google maps, rendering elevation, azimuth for any times, latitude, longitude, etc.: http://bit.ly/1kM2aM0
- Sunpath diagrams for each 1° of latitude for the Northern and Southern Hemisphere: http://bit.ly/1mCSJQv
- NASA cloud cover data: http://www.eso.org/gen-fac/pubs/astclim/espas/world/ION/ion-cloud.html
- NASA surface meteorology and solar energy (SSE) database of solar insolation, rain and wind data: http://eosweb.larc.nasa.gov/sse/
- PVGIS: solar radiation data for Europe, Africa and South-West Asia, and ambient temperature for Europe, plus terrain and land cover: http://re.jrc.ec.europa.eu/pvgis

- Solar path calculator; a solar spectrum calculator; an installed system cost calculator; and calculators for solar cell operation and future module prices: http://www.pvlighthouse.com.au/calculators/calculators.aspx
- TMY files: typical annual profiles of exterior climate data such as ambient temperatures, wind direction and velocity, precipitation, direct and diffuse irradiance (free) hourly climate data for many locations worldwide in the .epw format: https://energyplus.net/weather/sources
- World solar irradiation data: detailed maps, and maps of direct normal irradiation (free): http://solargis.info/doc/71
- Irradiation data by country: www.sealite.com.au/technical/solar_chart.php. The unit used is kWhm^{-2} day^{-1}
- Economic analysis of a photovoltaic system, with the determination of payback and chart: http://bit.ly/1j3pS3C
- Unit of measure converter: http://bit.ly/1iXDzqP
- Degree days: degreedays.net; US data: Climate Predication Center: www.cpc.noaa.gov/products/monitoring_and_data; Canadian Integrated Mapping and Assessment Project: http://bit.ly/2tLyVHo

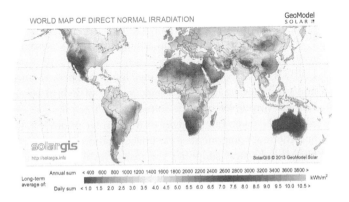

Figure 11.1 SolarGIS © 2014 GeoModel Solar. Reproduced with permission

11.2   Software

- The US Department of Energy (DOE) Building Energy Software Tools Directory, a comprehensive list of all simulation programs available: https://buildingdata.energy.gov/cbrd/resource/705

- EnergyPlus models heating, cooling, lighting, ventilating and other energy flows as well as water in buildings: https://energyplus.net/

- DASTPVPS (Design and Simulation Tool for PhotoVoltaic Pumping Systems): www.ibom.de/dastpvps.htm

- IDA ICE: building simulation software for predicting and optimising heating loads, cooling loads, energy consumption and thermal comfort in buildings: www.equa.se

- Polysun: for the prediction of system profit ratio (early planning phase) and system optimisation (detail planning); has a full featured solar simulation engine, including in-depth representation of thermodynamic and photovoltaic physics; offers the combination of solar thermal, photovoltaic and heat pump applications and enables easy PV system sizing. Calculations are based on a dynamic simulation model and statistical weather data: http://bit.ly/1sNfHHC

- PVsyst: for the study, sizing, simulation and data analysis of complete PV systems: www.pvsyst.com

- PV*SOL: for the design and simulation of PV systems: www.solardesign.co.uk

- HYBRID2 (USA): perform detailed long-term performance and economic analysis on a wide variety of hybrid power systems: http://photovoltaic-software.com/free.php

- SKELION: Sketchup's plugin to insert solar photovoltaics and thermal components in a surface: skelion.com

- HOMER Legacy: simplifies the task of evaluating design options for both off-grid and grid-connected power systems for remote, stand-alone and distributed generation (DG) applications: www.homerenergy.com/HOMER_legacy.html

- RETScreen: evaluates the energy production and savings, costs, emission reductions, financial viability and risk for various types of renewable energy and energy-efficient technologies: https://www.nrcan.gc.ca/energy/software-tools/7465
- RETScreen Software Solar Water Heating Model: https://www.nrcan.gc.ca/energy/software-tools/7465
- T*Sol: simulation for solar thermal heating systems: www.solardesign.co.uk
- Solar cooking: NASA's Surface Meteorology and Solar Energy Database can reveal if there is enough sunshine to use solar cooking in any part of the world. Levels > 3.9kWh/m^2 are suitable for solar cooking: www.eso.org/gen-fac/pubs/astclim/espas/world/ION/ion-pwv.html

11.3   Case studies and technical information

- International Energy Agency special projects:
  o   solar heating and cooling: www.iea-shc.org
  o   district heating and cooling: www.iea-dhc.org
  o   photovoltaics power system program: www.iea-pvps.org
  o   solar power and chemical energy systems: www.solar-paces.org
- ASHRAE (the American Society of Heating, Refrigerating and Air-Conditioning Engineers): www.ashrae.org
- SOLARGE, about solar thermal on buildings of all types, finished in 2007, but still containing useful resources: www.solarge.org

Table 11.1 Summary of providers of solar irradiation information

Product	Area	Period	Provider	Temperature resolution	Spatial resolution	Access	Price
NASA	World	1983–2005	NASA	average daily profile	100km	eosweb.larc.nasa.gov/sse	free
Meteonorm	World	1981–2000	Meteotest	synthetic hourly/min	1km (+SRTM)	www.meteonorm.com	on request
Solemi	World	1991–>	DLR	1h	1km	http://bit.ly/2gdLLdr	on request
Helioclim	World	1985–>	Mines-ParisTech	15min/30min	30km/3–7km	soda-pro.com	on request
EnMetSol	World	1995–>	University of Oldenburg	15min/1h	3–7km/1–3km	http://bit.ly/2gvycEA	on request
Satel-light	Europe	1996–2001	ENTPE	30min	5–7km	uib.no/en/rg/meten/54530/satel-light	free
PVGIS	Europe	1981–90	JRC	average daily profile	1km (+ SRTM)	photovoltaic-software.com/pvgis.php	free

Index